Thomas M. Turner

Fundamentals of Hydraulic Dredging

Second Edition

Historic gathering of three large U.S. Army Corps Hopper dredges—McFarland, Wheeler, and Essayons—to perform emergency dredging on the lower Mississippi in July 1995. Courtesy: U.S. Army Corps of Engineers.

Thomas M. Turner

Fundamentals of Hydraulic Dredging

Second Edition

Published by
ASCE Press
American Society of Civil Engineers
345 East 47th Street
New York, New York 10017-2398

ABSTRACT

This fully revised and updated edition presents the basic principles of hydraulic dredging in terms that are easily understood. It is non-theoretical and readable, in addition to being one of the most widely used texts available on dredging. The author, Thomas M. Turner, is a respected expert who has made substantial contributions to the field. The book is intended for dredge operators, along with government agencies and members of the legal profession who are concerned with the dredging industry. This updated edition includes new information on significant technical advances and environmental issues, and also includes both standard and metric units of measurement.

Library of Congress Cataloging-in-Publication Data

Turner, Thomas M., 1922-
    Fundamentals of hydraulic dredging / Thomas M. Turner. — 2nd ed.
      p.   cm.
    Includes bibliographical references and index.
    ISBN 0-7844-0147-0
    1. Dredging. 2. Dredges. I. Title.
    TC 187.T87 1996   95-53878
    627'.73—dc20   CIP

       The material presented in this publication has been prepared in accordance with generally recognized engineering principles and practices, and is for general information only. This information should not be used without first securing competent advice with respect to its suitability for any general or specific application.
       The contents of this publication are not intended to be and should not be construed to be a standard of the American Society of Civil Engineers (ASCE) and are not intended for use as a reference in purchase specifications, contracts, regulations, statutes, or any other legal document.
       No reference made in this publication to any specific method, product, process or service constitutes or implies an endorsement, recommendation, or warranty thereof by ASCE.
       ASCE makes no representation or warranty of any kind, whether express or implied, concerning the accuracy, completeness, suitability or utility of any information, apparatus, product, or process discussed in this publication, and assumes no liability therefore.
       Anyone utilizing this information assumes all liability arising from such use, including but not limited to infringement of any patent or patents.

Photocopies. Authorization to photocopy material for internal or personal use under circumstances not falling within the fair use provisions of the Copyright Act is granted by ASCE to libraries and other users registered with the Copyright Clearance Center (CCC) Transactional Reporting Service, provided that the base fee of $4.00 per article plus $.25 per page is paid directly to CCC, 222 Rosewood Drive, Danvers, MA 01923. The identification for ASCE Books is 0-7844-0147-0/96 $4.00 + $.25 per page. Requests for special permission or bulk copying should be addressed to Permissions & Copyright Dept., ASCE.

Copyright © 1996 by the American Society of Civil Engineers,
All Rights Reserved.
Library of Congress Catalog Card No: 95-53878
ISBN 0-7844-0147-0
Manufactured in the United States of America.

First edition published by Cornell Maritime Press, Inc. 1984.

Printed on recycled paper. 85% recovered fiber and 15% post-consumer waste.

Dedication

to June
best friend, critic, advisor
& dear wife

ABOUT THE AUTHOR

Thomas M. Turner, an honors graduate in engineering from North Carolina State University, served as naval engineering officer aboard landing craft in World War II. Since then he has worked in engineering and development for the Procter and Gamble Company, as chief engineer of the Buckeye Cotton Oil Company, and as head of the equipment department for Charmin Paper. In 1966 Turner began work as vice president of engineering for the Ellicott Machine Corporation, the oldest and largest builders of dredges in the United States, and became general manager of the Ellicott Dredge Division. In 1978 he established his own consulting firm, specializing in dredge efficiency, trouble-shooting, training seminars, and litigation assistance. He is a contributor to the *Proceedings* of the World Dredging Conference and to other professional journals.

CONTENTS

PREFACE   xi

Part I: Theories of Dredging

1. HYDRAULICS SIMPLIFIED   3
   Introduction to Dredge Hydraulics—Velocity Head—Centrifugal Pump Principle—Pump Affinity Laws—Specific Gravity Effect—Slurry Effect—Pump Head Efficiency Coefficients

2. PRODUCTION RATE CALCULATION   15
   Dredge Law I—Production Equation—Porosity, Weights, Specific Gravity—Production Measurement Systems—Volume to Weight Conversion—Grams per Liter Conversion

3. DREDGE EFFICIENCY   25
   Dredge Law II—Maximum Percent Solids Vs. Cavitation—Operating Time Vs. Downtime—Dredge Efficiency Chart—Swing Width Effect—Definition and Calculation—Dredge Law I Rephrased

4. HYDRAULIC TRANSPORT FACTORS   32
   Dredge Law III—Turbulence Requirement for Hydraulic Transport—Suction Velocity—Turbulence Requirements of Different Materials—Velocity Requirements of Different Pipe Sizes—Soil Classification

5. MAXIMUM DREDGE PRODUCTION 41
    Dredge Law IV—Barometric Head Induces Flow—
    Dredge Flow Varies with Suction Line Velocity and
    Area—Effect of Altitude on Velocity—Effect of Altitude
    and Temperature on Horsepower

6. THE SUCTION LINE AND DIGGING DEPTH 49
    Dredge Law V—Analysis of Suction Line Losses—Velocity Head—Entrance Loss—Friction Loss—Specific
    Gravity Head—Suction Lift—Optimizing Suction Velocity

7. HORSEPOWER VS. LINE LENGTH 56
    Dredge Law VI—Horsepower Vs. GPM, SG, and h—
    Pipeline Size Vs. Friction—h_F Vs. GPM and Pipe
    Diameter—Effect of Suction Size on Pumping Distance—
    Horsepower Vs. Line Size (Horsepower Coefficient)—
    Recommended Pump Horsepower

8. PRODUCTION CHARTS 69
    Dredge Law VII—Barometric, Torque, and Velocity
    Limitations on Production—Suction Line Size—Discharge Line Size—Booster Pump Effect—Ladder Pump
    Effect

9. THE DREDGE CYCLE 82
    Head-Capacity Curve on Water—System Resistance on
    Water—Head-Capacity Curve on Slurry—System Resistance on Slurry—Dredge Cycle Explained

10. FLOW REGIME AND FRICTION 87
    Friction Head Losses—Flow Regimes—Soil Types—
    Hazen-Williams Equation—Friction—Factor Chart

11. CAVITATION: CAUSES AND AVOIDANCE 94
    Definition—Cavitation Chart—Eye Speed—NPSHR—
    Impeller Geometry and Speed

Part II: Dredging in Practice

12. SELECTING THE DREDGE TYPE 101
    Plain Suction—Trailing Suction—Cutterhead—Compensated Cutterhead Dredge

CONTENTS ix

13. THE CUTTER 113
    Types and Functions—The Basket Cutter—The Bucket Wheel—The Endless Chain—The High Speed Disc Cutter

14. THE DREDGE PUMP 137
    Pump Type—Particle Clearance—Fully Lined Vs. Partially Lined Pump—Impeller—Stuffing Box—Shaft and Bearings—Adjustable Mounting—Wiper Vanes—Casing—Eye Speed—Tip Speed—Eye Diameter Vs. Impeller Diameter—Horsepower Coefficient—Drive—Torsional Vibration—Thrust

15. LADDER AND BOOSTER PUMPS 159
    Ladder or Submerged Pump—Design Requirements for Ladder Pumps—Suction Jet Booster—Natural Gas Problems—Ladder Pump Drives—Booster Pumps Vs. Transport Distance—Coordination with Dredge Pump—Water Hammer—Location of Booster

16. WEAR IN PUMPS AND PIPELINES 173
    Life Vs. Wear—Life Equation—The K Factor—Hydraulic Design Factor—The Brinnell Hardness Factor—Solids Concentration—Pump Size—Velocity—Weight of Solids—Particle Size—Angularity—Application of the Life Equation—Simplified Pipeline Life Equation—Normal Range Validity—Pressure—Corrosion—Wear Zones

17. AUXILIARY EQUIPMENT 185
    Forward Winch—Swing Speed—Line Pull—Anchors—Anchor Booms—Spud Hoist Winches—Spuds—Spud Carriage—Wire Rope

18. INSTRUMENTATION AND AUTOMATIC CONTROL 199
    Definitions—Durability—Accuracy—Dredge Position—Slurry System—Dredge Cycle Automation—Cutter Module—Digging Depth—Sounding—Winches

19. CALCULATING AND BIDDING THE PROJECT 208
    Contract Document Evaluation—Method of Calculation—Material to Be Pumped—Digging Depth—Terminal Elevation—Discharge Line Length—Cutter Capability—Height of Work Face—Swing Width—Type of Advancing Mechanism—Dredge Efficiency—Suction Line Size—Hourly Production Rate Vs. Production Time—Total Yards—Production Time—Calendar Time—Trash Vs. Production Time—Costs—Bid Price

20.	THE PERSONAL COMPUTER IN DREDGE MANAGEMENT Need—Software—Data Base—Accuracy—"D" Vs. "L" Dredges—PC Program Output—Simulation—Training—Computer Advantages	220
21.	OPERATION AND TROUBLESHOOTING Operational Errors and How to Avoid Them—Swing Angle for Advance—Swing Width—Anchor Location—Channel Width Limitations—Troubleshooting—Abnormal Gauge Readings and Their Meaning	229
22.	THE ENVIRONMENT AND THE DREDGE Environmentalists Vs. Developers—Water Pollution Defined—Politics and Public Opinion—Turbidity—Dredge As Cleanup Tool—Efficiency Vs. Environmental Disruption—Recommendations	239

ABBREVIATIONS	245
USEFUL FORMULAS AND CONVERSION FACTORS	247
REFERENCES	249
PUBLIC LAW 95-269	250
INDEX	255

PREFACE

This book is intended as a handbook for all involved with hydraulic dredges, be they owners, managers, engineers, levermen, or lawyers. It is simply phrased, and its fundamental truths are intended to guide the way to efficient, economic operation of the hydraulic dredge; it is also potentially helpful to the resolution of contract disputes, including litigation.

The Basic Dredge Laws of the first book are repeated in this second edition. As fundamental truths, they have stood the test of time, and provide a sound foundation on which to build one's knowledge of hydraulic dredge rheology. The charts and data from the first edition have been updated, improved, and simplified, but the fundamentals, while embellished, remain unchanged. An understanding of these fundamentals can steer the dredge operator away from the erroneous beliefs and practices of the past. The industry is still afflicted by some persistent myths, which hopefully this book will help to dispel. One myth, accepted for decades, is that solids suspended in liquids can not absorb, store, or transmit pressure energy. The Texas A&M University Center for Dredging Studies demonstrated the error of this statement with a simple laboratory apparatus. A tall, transparent cylinder filled with a sand and water mixture indicated a static pressure equal to the water height when the sand was motionless at the bottom of the cylinder; however, when the cylinder was inverted, the static pressure registered a higher value until the sand again settled on the bottom. Suspended solids can obviously convert their kinetic energy into pressure, although the conversion is not as efficient as that of a true (Newtonian) fluid.

The preface of the first edition of this book (1984) commented on the scarcity of technical literature regarding the subject of hydraulic dredging. Today, this situation is much improved. While the industry's authors are not yet prolific, the World Dredging Association, the Texas A&M University Center for Dredging Studies, the American Society of Civil Engineers and others have succeeded in encouraging publications that have contributed significantly to the technical progress of this essential industry; however, an appraisal of the industry's equipment would undoubtedly reveal that much of it lags behind the literature. Old inefficient designs remain, component configurations are inappropriate, and few prime movers are matched properly with their dredge pumps. Obviously, the education of the dredging industry needs to continue.

There are reasons for the industry's shortcomings. First, it is small, without the financial clout to support the fundamental R&D effort required to keep up-to-date in this fast-changing, technical world. The U.S. Army Corps of Engineers, through its developmental arm, the Waterways Experiment Station (WES) in Vicksburg, Miss., spends substantial funds on dredging activities; however, WES is a public institution and necessarily responds to political pressures. Environmental concerns, highly politicized, have claimed the lion's share of WES's attention, leaving only minor resources for equipment development. The annual cost of the U.S. waterways' dredging program, as managed by the Corps, is directly related to the efficiency of the U.S. dredging industry. Army Corps development improving this efficiency could lead to a reduction in the nation's dredging costs and thus justify itself. There are several development projects worthy of WES's attention.

In addition to WES and the Center for Dredging Studies, other organizations have contributed to the industry's knowledge. Various equipment manufacturers have helped—for example, GIW Industries with its large pump-testing facility, and Ellicott Machine Corporation with its versatile dredge-simulation facility. There is information in this book that has been derived from each of the four organizations mentioned, but there is still much to do to meet the needs of the industry. Frequently we must "make do" with the best information available, even when we question its adequacy. The author has included charts in this book for which he would like additional data. For example, the Chapter 1 chart showing the coefficient for pump-head efficiency (Che) and the Chapter 13 chart showing cutter capacity have been developed from limited data. They have been used with success, but the author would gratefully accept confirming or corrective data from other sources. Even though the charts are not fully validated in their present form, they represent much better data than was previously available to the author.

The industry has been slow in utilizing the personal computer (PC) to optimize the design and operation of hydraulic dredges. Seldom has there been an industry more needful of the capabilities of the PC. When one considers the project variables encountered by the hydraulic dredge (channel width, depth, line length, terminal elevation, dredged material characteristics), plus the dredge design variables (suction and discharge line size, pump RPM and horsepower, tip speed, eye speed, advance mechanism), the numerous possible combinations become apparent, and the need for a PC with appropriate software becomes obvious. It is hard to overemphasize the value of having an electronic simulation of a dredge available on the manager's desk for analysis of any project. Prompt answers are available to questions such as "What effect will adding 40 feet of terminal elevation to the discharge line have on production rate?" or "What effect will changing digging depth (or line length, or dredged materials) have on production costs?" The inability to answer such questions accurately has plagued the industry, leading to failed projects and bankrupt companies. Hopefully this book will encourage dredge operators to utilize this modern tool; those that do not will surely not be able to compete with those that do.

There are still newcomers to the industry who underestimate the complexity of dredge hydraulics. Operators who feel the dredge is "just a barge with a pump on it" contribute to the high bankruptcy rate in the industry. They observe a dredge and see the hull, winches, spuds, cutter, and pump, and conclude the dredge is no more complicated than much of the equipment used in the construction industry. This observation overlooks the complexity of the primary function of the dredge: the hydraulic transport of solids. This book is devoted largely to the explanation of that complicated, multifaceted process that goes on inside the dredge pump and the opaque dredge pipe. It is intended to make available to dredge operators not only the fundamentals of hydraulic dredging, but also a practical approach to a successful, economical operation.

A two-step program to assure a company's technical success with its hydraulic dredges is to (1) disseminate the fundamentals of this book throughout the company; and (2) utilize a personal computer with effective software to simulate the characteristics of each company dredge and each project parameter. Checking the project conditions against the design characteristics of the dredge with a desktop computer can lead the way to a successful field operation, avoiding the disastrous economic results of an inadequate dredge on a difficult project.

The second edition of this book, unlike the first, uses both the British/American and metric systems of measurement. A European critic wrote a complimentary review of the first edition, but pronounced it unfortunate that the book used only the B/A system. The American industry has not yet completed its conversion to the International System (SI) of units; however, a 1994 decree by the U.S. Army Corps of Engineers requires use of the SI system on future federal projects. Therefore, while this book uses the British/American system, it also provides parallel metric units to broaden its usefulness, both in the U.S. and abroad.

Public Law 95-269, passed by the U.S. Congress in 1978, represents a significant event in the history of the U.S. dredging industry. This act limited the in-house dredging fleet of the U.S. Army Corps of Engineers, as well as its direct dredging of public waters; in effect, it enfranchised the private industry to perform the Corps' routine dredging projects on the nation's waterways as long as the quoted price was no more than 25% above the Corps' estimate. The law has had a dramatic effect on the private industry. A substantial fleet of privately owned split-hull hopper dredges has been built, where none existed before. The law is reproduced as an appendix to this book.

One of the author's most rewarding experiences was the phone call received from an American dredge operator, who said, "I am calling to tell you that your book has not only educated us in dredging fundamentals, but we feel it has actually saved our company." If the second edition of *Fundamentals of Hydraulic Dredging* can continue in this mode, it will be as satisfying to the author as to the industry.

Part I
THEORIES OF DREDGING

Chapter 1

HYDRAULICS SIMPLIFIED

INTRODUCTION TO DREDGE HYDRAULICS

The study of the hydraulic dredge system involves an analysis of a unique application of fluid flow. Although the classic hydraulic principles apply, in practice, a dredge system is complicated by the presence of solids which make significant changes in the rheological, or flow, characteristics of the liquid.

Water, the normal dredge liquid, is a true (i.e., Newtonian) fluid which complies with the classic fluid flow principles. However, when solids are added in the form of sand or other sea-bottom material, water becomes a slurry with widely fluctuating flow characteristics, varying as a function of the type and percentage of the solids in it.

The student of hydraulic dredging must understand the unique nature of slurry flow as well as the classic fluid flow principles since a slurry system is actually a hydraulic transport system for solids. As applied to a hydraulic dredge, slurry flow is used to transport solids (sea-bottom material) from the cut or channel to the fill or deposit area.

The hydraulic dredge is a floating machine which removes sea-bottom material by entraining it in induced water flow, and transports it in a closed conduit to a designated deposit area. While there are several versions of the hydraulic dredge, e.g., plain suction, cutter head, hopper, and dust pan, they all comply with the principles outlined in this book.

VELOCITY HEAD, h_V

One of the most important concepts of dredge hydraulics for the dredgeman to understand fully is that of velocity head. As defined, velocity head is the vertical distance through which a liquid would have to fall to attain a given velocity, and as implied, it relates available head to resulting velocity or vice versa. Since velocity head, h_v, is involved in nearly every aspect of fluid flow calculations, whether it be entrance loss, pipeline friction, or pump design, it does require that the prospective dredgeman master the theory behind it as well as its practical applications. Velocity head is the basic building block in the calculation of any fluid flow system, and can be expressed as follows:

$$h_V = \frac{V^2}{2g} \qquad \qquad [\textit{Equation 1-1}]$$

Or: velocity head equals velocity squared divided by two times the acceleration of gravity.

This expression can be rearranged as follows:

$$V = \sqrt{2gh} \qquad \qquad [\textit{Equation 1-2}]$$

Here it is apparent that to induce a velocity of V, a head of h feet of a liquid is required, and is related to V as the square root of the product of h and two times the acceleration of gravity. Head, h_V (or pressure, if preferred), is expressed in feet of liquid being pumped. If the liquid is water with specific gravity (SG) of 1.0, the conversion to pounds per square inch (psi) is simple. One cubic foot of fresh water weighs 62.4 pounds. See Fig. 1-1. Obviously the height of the water column in the cube is one foot. Therefore, the head at the bottom of the cube is one foot. The area of the bottom of the cube is 12 inches times 12 inches which equals 144 square inches, and since the total weight of the water is 62.4 pounds, the

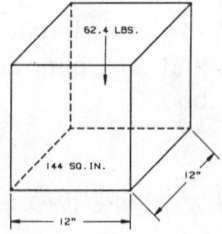

Fig. 1-1. Head, 1 cubic foot of fresh water.

HYDRAULICS SIMPLIFIED

unit pressure on the bottom of the cube is:

$$\frac{62.4 \text{ pounds}}{144 \text{ square inches}} = 0.433 \text{ psi}$$

Therefore, 1 foot of water head equals 0.433 psi, and 10 feet equals 4.33 psi.

If a fluid other than water is being pumped, the equivalent head in water is calculated by the simple multiplication of the fluid head by its specific gravity. For example, if a slurry of 1.4 specific gravity has a velocity head of 5 feet, the water equivalent is 5 × 1.4 = 7.0 feet. This is a very useful concept for the dredgeman since feet of water is easy to measure and conceive; also, the natural barometric head (the source of pump suction head) is 34 feet of water, which corresponds to 30 inches of mercury and 14.7 psi. The dredgeman should recognize that head is normally expressed in feet of liquid, and pressure in psi, but each is easily convertible into the other.

Velocity is expressed in feet per second, a commonly understood expression. But, not so commonly understood is acceleration which measures the *change* in velocity. If velocity is increasing by 1 foot per second each second, then the expression becomes 1 foot per second, per second. Although this may sound redundant to the layman, acceleration is expressed in *feet per second, per second* (ft/sec/sec) since velocity is increased by one foot per second every second.

The velocity head concept is so important to the dredgeman that the derivation of the expression $h_V = V^2/2g$ can provide worthwhile insights. Assume that a free-falling body falls from a head of h. See Fig. 1-2.

The acceleration of gravity at sea level, g, is measured as 32.2 ft/sec/sec. A falling body encountering negligible resistance will increase its velocity by 32.2 ft/sec every second it is falling freely. Expressed in equation form, this is:

$$V = gt \qquad [Equation\ 1\text{-}3]$$

Where V = the instantaneous velocity
g = the acceleration of gravity (32.2 ft/sec/sec)
t = time in seconds the body falls

Therefore, if the body starts at zero velocity, after one second it reaches 32.2 feet per second; after two seconds, 64.4 feet per second, and the table below can be extrapolated as far as desired.

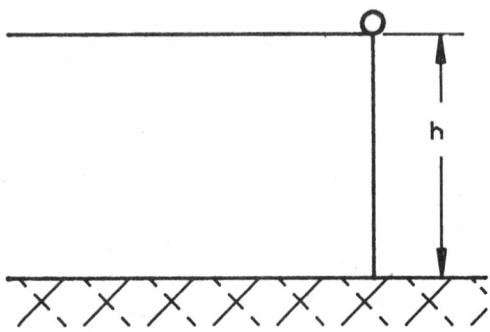

Fig. 1-2. Free-falling body.

	Instantaneous Velocity		Average Velocity	
Time-seconds	ft/sec	m/s	ft/sec	m/s
0	0	0	0	0
1	32.2	9.8	16.1	4.9
2	64.4	19.6	32.2	9.8
3	96.6	29.4	48.3	14.7
4	128.8	39.3	64.4	19.6

The velocity derived from Equation 1-2 is the instantaneous velocity achieved *after* the elapsed time. It is not the average velocity of the falling body which started at zero. With constant acceleration, the average velocity is the initial velocity zero plus the final, or instantaneous, velocity divided by two.

$$V_{average} = \frac{0 + V}{2} = \frac{V}{2}$$

In order to determine the distance, h, a body falls in the elapsed time, t, we multiply the average velocity by time.

$$h = \frac{V}{2} \times t \qquad [Equation\ 1\text{-}4]$$

So the distance, h, a body falls freely can be calculated simply by knowing the time it falls, using Equation 1-3 to obtain V and Equation 1-4 to obtain h. If we do not know time but do know either distance or velocity, we can calculate the unknown factor. From Equation 1-3 we know V = gt which, expressed in terms of time, becomes t = V/g. Now substituting V/g for t in Equation 1-4, we get:

$$h = \frac{V}{2} \times \frac{V}{g} = \frac{V^2}{2g} \qquad [Equation\ 1\text{-}1]$$

This is the classic velocity head expression. Expressed in terms of velocity it becomes:

$$V = \sqrt{2gh} \qquad [Equation\ 1\text{-}2]$$

a very useful equation for calculating velocity when only h is known.

What does this simple physics lesson have to do with hydraulic dredge principles? The dredgeman should know that a true fluid (called a Newtonian fluid in honor of Sir Isaac Newton who expounded the law of gravity) acts in accordance with the laws for a free-falling body. If, instead of a free-falling body as indicated in Fig. 1-2, we have a tank with a height of liquid, h, and an open nozzle at the bottom, Fig. 1-3, the fluid will flow through the nozzle in accordance with the velocity head expression, Equation 1-1 and Equation 1-2. Note that if h equals 4 feet in Fig. 1-3, the velocity at the nozzle will be:

$$V = \sqrt{2gh} = \sqrt{2 \times 32.2 \times 4} = 16 \text{ ft/sec}$$
$$= \sqrt{2 \times 9.8 \times 1.22} = 4.9 \text{ m/s}$$

Also, if we know that the velocity at the nozzle is 16 ft/sec, we can calculate the head creating the velocity as follows:

$$h = \frac{V^2}{2g} = \frac{(16)^2}{2 \times 32.2} = \frac{256}{64.4} = 4 \text{ ft or}$$
$$= \frac{(4.877)^2}{2 \times 9.8} = 1.22 \text{ m}$$

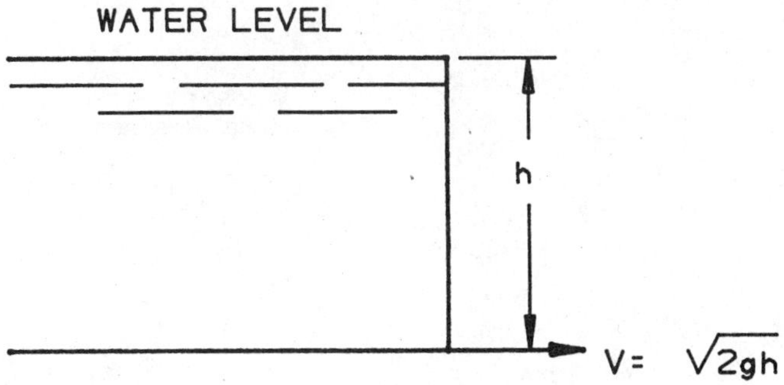

Fig. 1-3. Flow through nozzle.

The simple physical concept of velocity head is that it is the feet of head required to produce a given velocity. In the example above, 4 feet of head is required to produce a velocity of 16 feet per second.

Doubling the head available will not double the velocity since velocity varies as the square root of the head available. The following equations show that the effect on velocity when the head is first doubled and then quadrupled.

$$V = \sqrt{2gh} = \sqrt{2 \times 32.2 \times 8} = 22.7 \text{ ft/sec} = 6.9 \text{ m/s}$$
$$V = \sqrt{2gh} = \sqrt{2 \times 32.2 \times 16} = 32 \text{ ft/sec} = 9.75 \text{ m/s}$$

Note that when the head is quadrupled, in the second example above, the velocity is doubled. We can predict this readily by recognizing the square root relationship of h to V.

CENTRIFUGAL PUMP PRINCIPLE

For a Newtonian fluid, the effect of the velocity head equation is completely reversible. Not only will 4 feet of head create a velocity of 16 feet per second, but ignoring losses, 16 feet per second velocity entering through the nozzle as shown in Fig. 1-2 will create a head of 4 feet in the tank.

It is from the reversible nature of velocity head that the principle of the centrifugal dredge pump is derived. By rotating the vanes of the impeller through the fluid in the pump, the fluid is centrifugally impelled into the volute (casing) of the pump where most of the velocity is converted into pressure or head as a function of the velocity head expression. See Fig. 1-4.

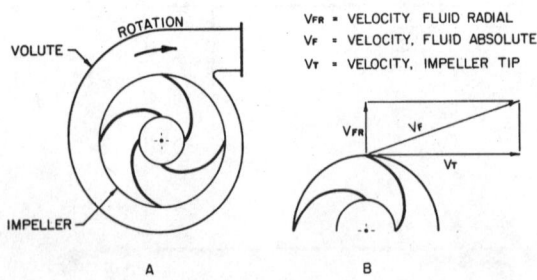

Fig. 1-4. Centrifugal pump with velocity components.

HYDRAULICS SIMPLIFIED

The remainder of the velocity energy imparted to the fluid is utilized to maintain the velocity through the pump and to overcome the flow losses through the pump. Surprisingly, the head achieved is generally something more than the velocity head calculated from the tip speed of the impeller. This is because the fluid achieves velocities higher than the impeller tip speed since the fluid must flow both circumferentially with the impeller and radially out into the volute, providing a resultant velocity higher than either component. See Fig. 1-4.

The head created by a pump has a fixed relationship to the flow rate through the pump at a given rotational speed. This relationship is generally determined by test and is graphically indicated by the head capacity chart or characteristic curve for the pump, Fig. 1-5. While at the rotational speed for which the curve is plotted, any change in capacity through the pump will result in a change in head, and vice versa. If either head or capacity is known along with rotational speed, the other factor can be determined from the head capacity chart since the pump must follow the fixed pattern described by the curve.

PUMP AFFINITY LAWS

Normally, the head capacity information is given for several speeds, along with horsepower and efficiency information. See Fig. 1-6. However, if only one speed is known, the performance of the pump at other speeds can be approximated by the use of the Pump Affinity Laws indicated as follows:

(1) $GPM \sim RPM$
(2) $\quad h \sim (RPM)^2$
(3) $\quad HP \sim (RPM)^3$

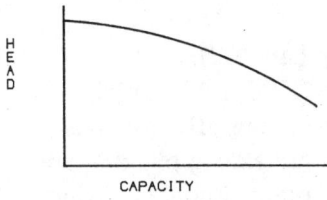

Fig. 1-5. Centrifugal pump head-capacity curve.

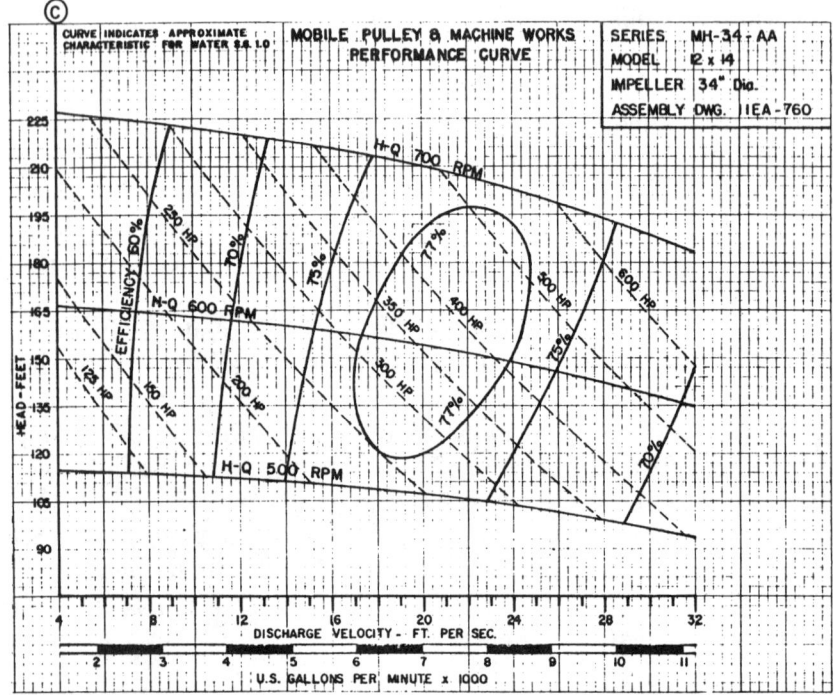

Fig. 1-6. Centrifugal pump head-capacity curve. Courtesy: Mobile Pulley & Machine Works.

In Pump Affinity Law 1, the GPM impelled into the volute and out the discharge nozzle increases in direct proportion to the RPM.

In Pump Affinity Law 2, the head increases as the square of the RPM or impeller tip speed. This reflects the velocity head relationship where V is impeller tip speed.

$$h = \frac{V^2}{2g}$$

In Pump Affinity Law 3, HP increases as the cube of the RPM. Since HP varies with GPM and h, which vary as the first and second power of RPM respectively, HP varies as the third power.

The affinity laws are also applicable when changing the outside diameter (OD) of a pump impeller, using tip speed as the determining factor.

SPECIFIC GRAVITY EFFECT

Curves for dredge pumps produced by their manufacturers are based on water for testing convenience and for comparison purposes. The purpose of the dredge pump, however, is to pump solids in slurry form, and the dredgeman should note the significant effects of replacing water, a true Newtonian fluid, with a slurry.

If we assume the slurry is composed of water and very fine solid particles so that the fluid performs essentially as a Newtonian fluid, then the major difference is specific gravity. Furthermore, if we assume the SG is 1.4 but that viscosity is the same as water, we can examine the pump head capacity curve and note that it does not change at all. However, the head capacity curve, which is identical to that of water, now represents feet of head of a 1.4 SG fluid rather than a 1.0 SG water. If a pressure gauge were on the discharge line, it would indicate approximately 1.4 times the pressure of water at the same flow rate. The horsepower curve would rise by 1.4 times that of water, whereas efficiency would not be affected, as long as the fluid was Newtonian.

This effect is easily detected in a dredge when the pump which has been pumping water suddenly receives a heavy slurry. A pressure increase is detected on the dredge discharge gauge, but of smaller magnitude than the 1.4 change in SG, because as the flow rate in the discharge line increases, the operating point on the head capacity curve moves to the right, reducing the head increase. For a further discussion of this effect, see Chapter 9, the Dredge Cycle.

SLURRY EFFECT

When solids are added to water forming a slurry, the SG is increased over that of water. Therefore, the HP and discharge pressure of the pump are increased, if other factors such as RPM do not change. Unfortunately, slurries in general are not true fluids and therefore do not comply fully with the Newtonian laws. The head and efficiency of the pump on slurries are reduced from that of water as a function of the nature and percentage of the volume of the solids in the slurry. The chart, Fig. 1-7, shows the coefficients to be applied to the water head and efficiency for various sized solids and percentage of slurries. Note that the *pressure* generated

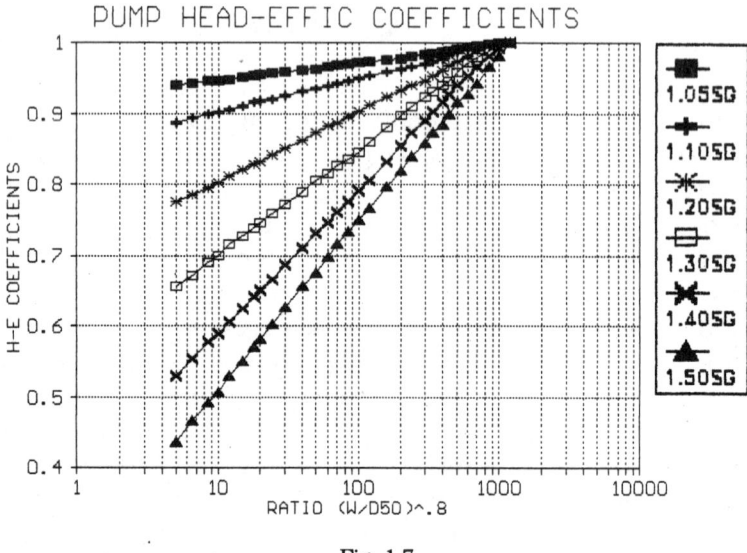

Fig. 1-7.

by a slurry is generally higher than for water in spite of the *head* reduction. However, the non-Newtonian slurry almost always makes a significant reduction in pump efficiency when compared to water.

PUMP HEAD EFFICIENCY COEFFICIENTS

As Fig. 1-7 discloses, the correction coefficient for water (where SG = 10) is always unity and no correction is needed. The industry has long recognized that the higher the solids concentration of the slurry and/or the coarser the solids material, the lower the head-efficiency coefficient (Che) will be. As the solids percentage and/or the median grain size (D_{50}) of a slurry is increased, the less the slurry acts like a Newtonian or true fluid.

There have been wide divergences in the values of Che within the industry. A comparison of test data on small slurry pumps (4 to 6 inches), with the empirical data for Che within the dredging industry (10 to 30 inches), has led to the explanation of the divergences. It is true that solids percent and grain size influence Che, but the magnitude of the effect varies broadly with pump size. For example, a particle 6.5 inches in diameter will not pass a 10-inch pump with a 6-inch passageway (distance between impeller shrouds W), but would easily pass a 30-inch pump with an 18-inch passageway. Similarly, a 5.5-inch particle will pass the 10-inch pump with difficulty (high losses), but will cause relatively minor losses when passing the 30-inch pump. Obviously, the median grain size is not the lone determining factor; however, it is a key factor in calculating the determining ratio of passageway size to that of the solids D_{50} $(W/D_{50})^{0.8}$.

Fig. 1-7 plots a function of the ratio against Che for several slurry specific gravities. The curves were plotted from valid but limited test data, and represent the first publication of this data other than the author's proprietary PC software. The author would like to see further data to fine-tune the chart; however, the chart has been used successfully for years, and is offered as a major improvement over the current, conflicting data of the industry.

SUMMARY

The material presented in this first chapter is intended to familiarize the reader with the broad principles of fluid flow, the centrifugal pump, and the effect of slurry on the pump performance. The next several chapters are devoted to specific hydraulic dredge principles.

Jean Rigal, the first American-built dredge with ladder pump and two dredge pumps. Courtesy: Ellicott Machine Corporation.

Chapter 2

PRODUCTION RATE CALCULATED

DREDGE LAW I
PRODUCTION VARIES AS FLOW TIMES AVERAGE PERCENT SOLIDS

The production of a hydraulic dredge is expressed as a quantity of *solids* transported. Water is the transport medium and is used to remove a *volume* of solids from a channel, or to fill a certain volume with solids, or both. The *weight* of the production is a function of its specific gravity, and is generally incidental to the project contract. However, since mining dredge operations deal in weight, weight cannot be ignored, and its effect on dredge production has to be considered.

The amount of water pumped is usually unimportant to the person or organization paying for the dredging. Nevertheless, it is important to the dredge operator who is not reimbursed for the cost of pumping water and therefore wishes to pump the minimum amount of water compatible with his production objectives.

PRODUCTION EQUATION

Equation 2-1 shows the simplest form of the production equation in the B/A system.

$$\text{cu yd/hr} = \text{GPM} \times \text{average percent solids} \times .297 \quad [Equation\ 2\text{-}1]$$

The solids percent is by in situ volume. The constant of .297 is the result of combining the factors for converting American GPM to cu yd/hr as follows:

$$GPM \times 60 \text{ min/hr}/(7.48 \text{ gal/cu ft} \times 27 \text{ cu ft/cu yd}) = .297$$

The following equation applies in the B/A system when calculating cubic yard per hour using the common dredge terms of inside diameter of pipe in inches, velocity in feet per second, and specific gravity of slurry:

$$\text{cu yd/hr} = d^2 \times v \times (SG-1) \times 661 \qquad [\textit{Equation 2-2}]$$

This is the convenient equation used by personal computers, whereas Equation 2-1 is simpler for discussion purposes.

Using Equation 2-1 or 2-2, production rate can be calculated if the GPM (or velocity) and average percent solids is known. Likewise, average percent solids, a very important measure of dredge efficiency, can be calculated if the velocity and production rate are known.

Dredge Law I is a fundamental statement that should become second nature to the dredgeman. Note that if the dredge is able to hold its GPM constant while doubling its percent solids, the production doubles. Likewise, if velocity can be doubled while holding percent solids constant, the production is doubled. If velocity is doubled and percent solids halved, production remains the same; however, if the velocity is halved *and* the percent solids halved, then production becomes only 25 percent of what it had been: $\frac{1}{2} \times \frac{1}{2} = \frac{1}{4}$ or 25 percent.

POROSITY, WEIGHTS, SPECIFIC GRAVITY

The cubic yards per hour in Equation 2-1 is normally based on the in situ measurement or "as found" volume of material in the cut. If we assume that the material being pumped is a clean, granular, silicate sand with a specific gravity of 2.65, the volume of this sand as found in the cut or channel will be the same as when hydraulically deposited in the fill or deposit area, since clean sand has essentially a zero swell or shrink factor. However, if clay or other compacted material is involved, there is a possibility that when measured in a calibrated canister or in the fill, it will have expanded in volume by 10, 20, 30 percent or more. It is important to the dredging contract to define clearly whether the unit payment will be based on cut or fill.

PRODUCTION RATE CALCULATED

For the instructional purposes of this book, we will normally assume a clean sand, although material to be removed from a cut can vary in specific gravity from slightly above that of water to a dense material of 2.65 or higher with zero porosity. (Specific gravity is defined as the ratio of the weight of a given volume of material to that of water.) Recognition of the nature of the material to be dredged can make the difference between financial success and failure of the project. The cost of dredging can vary from cents per cubic yard for the light material to several dollars for the dense, hard material.

For a typical granular sand, the porosity approximates 33.3 percent. On the bottom of the waterway, the granules rest on one another, and the interstices are filled with water. When the sand is pumped to the fill, the water may drain out, and the interstices may be filled with air; but the granules still rest on one another occupying the same space. This means that the weight of a cubic foot of water-saturated sand is approximately 21 pounds heavier than a dry cubic foot since the weight of the water is:

0.333 cu ft × 62.4 lbs/cu ft × SG of 1.0 = 20.8, say 21 lbs

Also, since the dry solids occupy the remaining 66.7 percent of the cubic foot, their weight is as follows:

0.667 cu ft × 62.4 lbs/cu ft × SG of 2.65 = 110.3, say 110 lbs

When completely wet, i.e., when the interstices are filled with water, the sand weighs 110 plus 21 of water for a total of 131 pounds. Logically then, a cubic foot of 2.65 SG sand will weigh anywhere from 110 to 131 pounds as a function of the amount of water it contains. At 131 lbs/cu ft

SG = 131/62.4 = 2.1 or 2100 g/l

PRODUCTION MEASUREMENT SYSTEMS

The metric calculation of the above is based on the simple concept of 1,000 parts in a liter [cubic centimeters (cc)], described in more detail at the end of the chapter.

Water: 333 cc × 1.0 SG = 333 g/l
Solids: 667 cc × 2.65 SG = 1,767 g/l
Totals: 1,000 cc × 2.1 SG = 2,100 g/l

Note that specific gravity (SG) and grams/liter (g/l) are ways of expressing the same weight data for the mixture. In the earlier B/A examples, the 2.1 SG is the same value as 2,100 g/l. The metric expression is numerically 1,000 times greater than the B/A SG.

Fig. 2-1 discloses the composition and weight of one cubic foot of slurry for water and a 2.65 SG granular material, varying from zero percent solids to 100 percent true volume. The formula is as follows:

$$\text{In situ SG} = \text{SG solids} \times \text{percent} + \text{SG water} \times (1 - \text{percent})$$

Where percent = percent solids by true volume, and in situ SG = SG of mixture.

Note in Fig. 2-1 that the 1.5 in situ SG (including the water content) is the practical maximum percent for hydraulic transport. The author has observed slurries as high as 1.6 SG, but the flow regime is unstable and the operator is unable to maintain it; thus flowing slurries of 2.65 solids are considered to vary from 1.01 to 1.5 SG. Note that the optimum 1.5 SG slurry consists of more solids by weight than water.

The 2.1 SG is the common condition assumed for 2.65 sands in their in situ, undredged condition. An in situ SG less than 2.1 suggests a material with organics, muds, or clays with attached water molecules. Such materials are not uncommon on maintenance jobs, but require a different production calculation technique reflecting the lower initial in situ SG.

The 2.65 SG represents the non-granulated, non-porous material that would normally be blasted before dredging. So, the operating dredge would not encounter pump slurries between 1.6 and 2.1, nor pump in situ materials above 2.1.

SPEC GRAV MIXTURE	TRUE VOL % SOLIDS	TRUE VOL % WATER	INSITU VOL % SOLIDS	MIX WGT LBS/CU FT	SOLIDS LBS/CU FT	WATER LBS/CU FT	GMS/LTR SOLIDS
2.65	100.00	0.00	150.00	165.4	165.4	0.0	2650
2.1	66.67	33.33	100.00	131.0	110.2	20.8	1767
2.0	60.61	39.39	90.91	124.8	100.2	24.6	1606
1.9	54.55	45.45	81.82	119.6	90.2	29.4	1445
1.8	48.48	51.52	72.73	112.3	80.2	32.1	1285
1.7	42.42	57.58	63.64	106.1	70.2	35.9	1124
1.6	36.36	63.64	54.55	99.8	60.1	39.7	964
1.5	30.30	69.70	45.45	93.6	50.1	43.5	803
1.4	24.24	75.76	36.36	87.4	40.1	47.3	642
1.3	18.18	81.82	27.27	81.1	30.1	51.1	482
1.25	15.15	84.85	22.73	78.0	25.1	52.9	402
1.2	12.12	87.88	18.18	74.9	20.0	54.8	321
1.1	6.06	93.94	9.09	68.6	10.0	58.6	161
1	0.00	100.00	0.00	62.4	0.0	62.4	0

Fig. 2-1.

PRODUCTION RATE CALCULATED

In situ volume is the volume as found or transported by the dredge. It has 33.3 percent voids, and thus normally is found with a 33.3 percent water volume. The granulated material occupies the same volume whether the voids are filled with water or air, so this is the volume with which the dredge project is concerned in both cut and fill. *All contractual matters deal with the measurable in situ volume.* True volume of a granulated material, i.e., with zero voids, does not occur in nature, and is not measurable. In situ volume is 150 percent of the theoretical true volume. The volume of solid rock, when blasted for dredging, becomes 50 percent greater as shown in the first row of Fig. 2-2. All calculations and formulas in this book deal with in situ volume unless indicated otherwise.

Weight of the dredged material is seldom of contractual significance to the dredgeman since he is normally paid by volume. Since weight can affect operational results, however, the subject will be discussed in Chapter 6.

Mining companies and academicians frequently deal in weight, generally expressed in short tons (2,000 pounds), long or metric tons (2,204.6 pounds), or gross tons (2,240 pounds). Most dredgemen use 1.5 short tons or 3,000 pounds as the weight of a cubic yard of sand. This is a reasonable approximation since the dry weight of a cubic foot of 2.65 SG sand equals 2,978 pounds (110.3 pounds × 27 cu ft/cu yd).

Following are four convenient production equations, two in the B/A system and two in the metric system. Note that the *average* solids content is used in the mathematical examples that follow the equations. All examples are for a 20 inch (508 mm) dredge, with a 15.12 feet per second velocity (14,805 GPM or 4.61 m/s), slurry SG of 1.25 (.2273 percent solids by volume or 402 g/l), producing 999 cu yd/hr (764 m³/hr).

cu yd/hr = GPM × volume percent solids/100 × .297
   = 14,805 × .2273 × .297 = 999   *[Equation 2-1 B/A]*
cu yd/hr = d^2 × velocity × (SG-1) × .661
   = 20^2 × 15.12 × .25 × .661 = 999   *[Equation 2-2 B/A]*
   m³/hr = m³/sec × g/l × 3.27
   = .9341 × 250 × 3.27 = 764   *[Equation 2-3 M]*
   m³/hr = d^2 × m/s × g/l × 2.57
   = $(.508)^2$ × 4.61 × 250 × 2.57 = 764   *[Equation 2-4 M]*

A graphical representation of the relationship between slurries expressed in in situ volume, true volume, and weight is shown in Fig. 2-2. A study of this graph is recommended in order to comprehend the significant differences in the specific gravities of slurry as expressed in terms of percent solids by weight, in situ volume, or in true volume. The dredgeman almost always uses in situ volume, but in mining operations weight may be used. True volume has little or no significance to the dredgeman since it implies zero voids which in turn implies large nonporous masses, not adaptable to hydraulic transport.

Note that in any of the production equations, the percent solids is always expressed in terms of in situ volume for clean sand, or any material with the same volume after pumping as before. Volume is the only physical dimension practical for the dredgeman to measure. If a typical sample could be obtained from a flowing slurry and placed in a calibrated canister, the sand would settle to the bottom and the percent volume could be read at the surface of the sand. This is the actual volume the sand will occupy when deposited in the fill, and represents a good indication of the solids percentage for calculation purposes. Note again that there are 33.3 percent voids in the sand and the in situ, or dredged volume is only

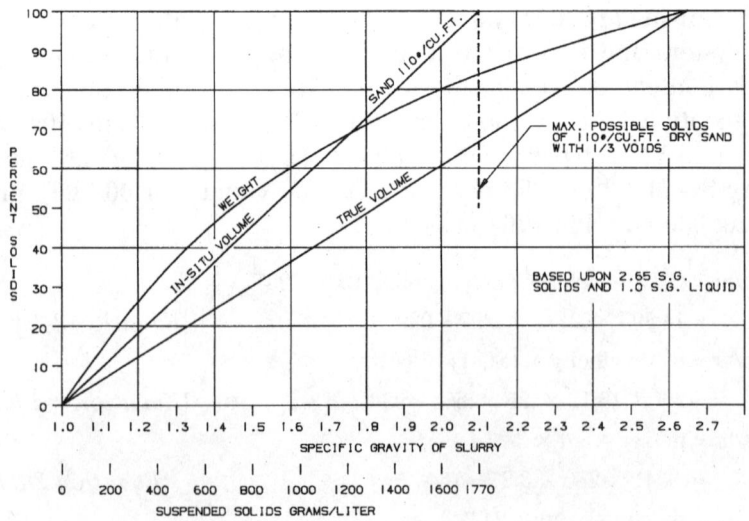

Fig. 2-2. Specific gravity of sand-water slurry vs. percent solids concentration.

66.7 percent of the "true" volume. (See Fig. 2-2.) It should be noted that this in situ volume is the actual volume measured by the dredgeman and is normally the basis upon which he is paid.

While clean, nonexpanding, nonshrinking sand will normally be used for instruction purposes in this book, it behooves the dredgeman to understand that some of the materials he encounters will expand or shrink when displaced from their in situ location and pumped to a fill area. When expansion occurs, the production of a dredge in the fill may appear to be greater than that measured in the cut. For example, if the material has a 30 percent expansion factor, the production as measured in the fill and calculated from the disturbed percent volume of a *sample in a graduated cylinder* will coincide. However, if the contract is being paid on the basis of in situ material removed from the cut, then the production equation will be affected as follows:

$$\text{Prod} = \frac{\text{Flow} \times \text{average percent solids}}{1.3}$$

Or: approximately 23 percent less than actually measured in the fill.

The converse is true in the case of shrinkage.

VOLUME TO WEIGHT CONVERSION

The following are formulas for determining slurry specific gravity when either the volume or weight fraction of solids is known:

$$C_V = \frac{SG_S - SG_W}{SG_{SO} - SG_W}$$

Where C_V = fraction of solids by volume
SG_{SO} = specific gravity of solids
SG_S = specific gravity of slurry
SG_W = specific gravity of liquid

For cold water, $SG_W = 1.0$, the formula can be stated as:

$$C_V = \frac{SG_S - 1}{SG_{SO} - 1}$$

When the concentration of solids is given by weight fraction C_W, i.e., weight of solids/weight of mixture, then:

$$C_W = \frac{SG_{SO}(SG_S - 1)}{SG_S(SG_{SO} - 1)} = \frac{C_V SG_{SO}}{SG_S}$$

$$C_W SG_S = C_V SG_{SO}$$

Solving for SG_S we get:

$$SG_S = \frac{(1 - C_V)}{(1 - C_W)}$$

Note that C_V is the fraction of solids expressed in terms of *true* volume, i.e., with no voids. To the dredge operator, who is paid for the volume of the sand on the bank including voids, *true* volume is an artificial concept which does not occur in nature. However, the classical formulas must include *true* volume for universal application, since percent voids varies broadly for different materials.

GRAMS PER LITER CONVERSION

The Waterways Experiment Station at Vicksburg, Mississippi, has adopted the metric system, grams per liter (g/l), for the expression of turbidity or suspended solids. The less definitive Jackson Turbidity Unit (Chapter 19) will eventually be replaced by grams per liter, so that the following relationships (courtesy of the Waterways Experiment Station) will be useful.[11]

Relationship between Suspended Solids Concentration, Bulk Density, and Percent Solids by Weight

Suspended solids concentrations (in grams per liter) can be converted to percent solids by weight or bulk density using the following procedure:

(a) $\dfrac{\text{Solids concentration (i.e., weight of dry solids)}}{\text{dry density of solids}}$ = volume of solids

(b) 1,000 cc of suspension − volume of solids = volume of liquid

(c) Volume of liquid × density of liquid = weight of liquid

(d) $\dfrac{\text{Weight of solids} \times 100}{\text{weight of solids} + \text{weight of liquid}}$ = percent solids (by weight)

(e) $\dfrac{\text{Weight of solids} + \text{weight of liquid}}{1,000 \text{ cc of sample}}$ = bulk density of sample (g/cc)

PRODUCTION RATE CALCULATED   23

Fig. 2-3. Concept of 24-inch offshore dredge with work barge and quarters barge. Courtesy: Ellicott Machine Corporation.

Example: solids concentration = 200 g/l,
density of solids = 2.65 g/cc
density of liquid = 1.03 g/cc

(a) $\dfrac{200 \text{ g}}{2.65 \text{ g/cc}} = 75.47$ cc of solids

(b) 1000 cc − 75.47 cc = 924.53 cc of liquid

(c) 924.52 × 1.03 g/cc = 952.27 g of water

Where: density of fresh water = 1.00 g/cc
density of seawater = 1.035 g/cc

(d) $\dfrac{200 \text{ g} \times 100}{200 \text{ g} + 952.27 \text{ g}} = 17.35$ percent solids by weight

(e) $\dfrac{200 \text{ g} + 952.27 \text{ g}}{1{,}000 \text{ cc}} = 1.152$ g/cc

Note that seawater grams per liter is used in the example, and *true* volume rather than in situ.

Parts per million (ppm) = mg/l (milligrams/liter)
Parts per thousand (ppt) = g/l (grams/liter)

SUMMARY

The dredge is a tool, and production, i.e., the removal of subaqueous solids, is its purpose. The simple production equation, regardless of units, underlies all discussion of dredge capacity, and is embellished further in Chapter 3.

Chapter 3

DREDGE EFFICIENCY

DREDGE LAW II

**AVERAGE PERCENT SOLIDS EQUALS
MAXIMUM PERCENT SOLIDS
TIMES DREDGE EFFICIENCY**

As shown in Chapter 2, Dredge Law I indicates that production is directly proportional to average percent solids. Dredge Law II introduces two new and important terms which require definition.

MAXIMUM PERCENT SOLIDS VS. CAVITATION

The maximum percent solids is defined as the highest practical, instantaneous percent of solids the hydraulic system can transport, without cavitating the pump.

Maximum percent solids occurs at a point easily recognizable by the dredge leverman. On a conventional hydraulic dredge, i.e., with a pump in the hull but no submerged pump, the leverman sets his pump speed to achieve the desired velocity in the pipeline. This results in perhaps a 7- to 10-inch mercury reading on the vacuum gauge, representing the head or pressure losses in the suction line on water alone. Then as the operator lowers his cutter (excavator) into the bottom material, the vacuum rises abruptly because of the demand of the heavy solids. At some point, perhaps 24 to 27 inches mercury, the pump becomes noisy, vibrates, and loses its pumping effectiveness because of a phenomenon called *cavitation*. (See Chapter 11 for a discussion of cavitation.) This occurs when the natural barometric pressure is no longer capable of overcoming the losses in the suction line at the rate the pump demands. The careful operator will control the pickup of solids so as to keep his vacuum indication an inch or two of mercury below the cavitation point. This, then, is the maximum percent solids which coincides with the maximum instantaneous production rate.

DREDGE EFFICIENCY DEFINED

By a simple rearrangement of Dredge Law II we see that:

$$\text{Dredge efficiency} = \frac{\text{average percent solids}}{\text{maximum percent solids}}$$

It is now apparent that dredge efficiency, D.E., is merely the ratio of the percent solids that the dredge *averages* over a period of time to the maximum practical percent solids achievable on an instantaneous basis.

Dredge efficiency and maximum percent solids are very important to the understanding of the hydraulic principles involved in dredging. Prior to the SG meter (see Chapter 18) the dredgeman had no real grasp of the *maximum* percent solids since he could only calculate *average* percent solids after measuring his production in the cut or fill by using the Dredge Law I equation. Since cavitation occurs on maximum percent solids, not average percent solids, many miscalculations have been made on suction line sizes.

Dredge efficiency is affected by many factors—operator skill, method of advancing the dredge, height of the submerged bank, etc. Fig. 3-1 shows the method of advancing for a conventional walking spud dredge. Also shown is the production diagram during a single swing and return cycle.

Referring to Fig. 3-1 note that at point A the dredge has swung the prescribed amount to the port (left) for advancing, after having completed the previous dredging swing. The operator, at this point, drops his walking (starboard) spud and raises his working (port) spud. The dredge now swings to point B, where the cutter has advanced its full length into the bank. (The advance can vary with any percentage of the cutter length as a function of the material being dredged.) The operator now lowers the working spud and raises the walking spud. The working spud is now a cutter length ahead of its previous position, and the dredge is on course.

From zero production at A, the dredge advances to 100 percent production at B, and continues at 100 percent to C. At C, however,

DREDGE EFFICIENCY

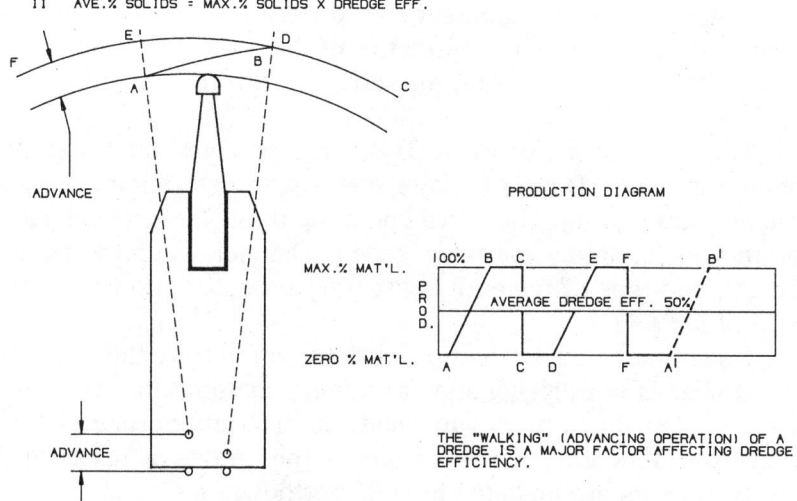

Fig. 3-1. Production while advancing.

the operator reverses the swing and the production is zero until point D (old point B) where the production starts to rise, reaching 100 percent at point E. It remains at 100 percent to point F, and then is zero when the swing is reversed to point A' (old E) where the advancing cycle is repeated.

An examination of the shape of the production diagram in Fig. 3-1 discloses that the dredge efficiency of a walking spud dredge approximates 50 percent; however, it is entirely possible for such a dredge to vary between 5 and 75 percent as a function of the dredging conditions and operator skill. Maintenance dredging, where the operator must "chase" the material, is generally low efficiency. River dredging for fill material, with no prescribed channel to dictate cutter position, can result in quite high efficiency. It behooves the dredgeman to understand the conditions that affect dredge efficiency and to estimate accordingly, notwithstanding the dredge manufacturer's production rating which is probably based on 50 percent dredge efficiency, and which can be demonstrated under the proper conditions. The author suggests that 40 percent D.E. is a better average in today's context, but the correct D.E. should always be calculated, not assumed, where data is available.

It should now be apparent that dredge production is directly proportional to D.E. The importance of the estimator's understanding of D.E., and the factors that affect it, are paramount to good results.

The distinction between D.E. and percent operating time should be clear. D.E. is a major element in determining the production rate during productive operating time. Factors such as operating skill, dredge advance speed, channel width, work face height, and type of material being dredged enter into the calculation of D.E.

Downtime is an expression of the hours during the day when the dredge is not digging and/or pumping material for any reason, whether scheduled or unscheduled. Routine and normal interruptions to production such as advancing the dredge or resetting the swing anchors are included in D.E., not downtime.

Operators discovered they could minimize the time it took to advance a dredge by increasing the spud and swing wire speeds to accelerate dredge advances. This is a small increase and is seldom singled out as a separate calculation for D.E.; however, any change that decreases non-productive time increases D.E. Installation of a spud carriage to replace the walking spud arrangement significantly decreases spudding time; therefore, it will increase D.E. and productivity by a percentage varying as a function of the job conditions.

The traditional 50 percent D.E. has encompassed wide variations in material type, e.g., standing material versus free-flowing materials, as well as work face heights that allow multiple swing cycles without the necessity of advancing. These factors, plus swing width, are significant enough to be considered separately in the calculation of D.E.

There has been little in the literature regarding the calculation of D.E. The author had the opportunity to participate in an experiment with an operating cutterhead dredge where the actual swing cycles were recorded, plotting slurry velocity and SG against time. The data were analyzed to determine D.E.; these results were then plotted and smoothed, yielding the curves of Fig. 3-2. (This is new data, not published previously except in the author's PC software.) The approximate center of the plot would reasonably support the use of 40 or 50 percent as a default value, but it should be noted that D.E. can vary from less than 5 percent to about 75 percent,

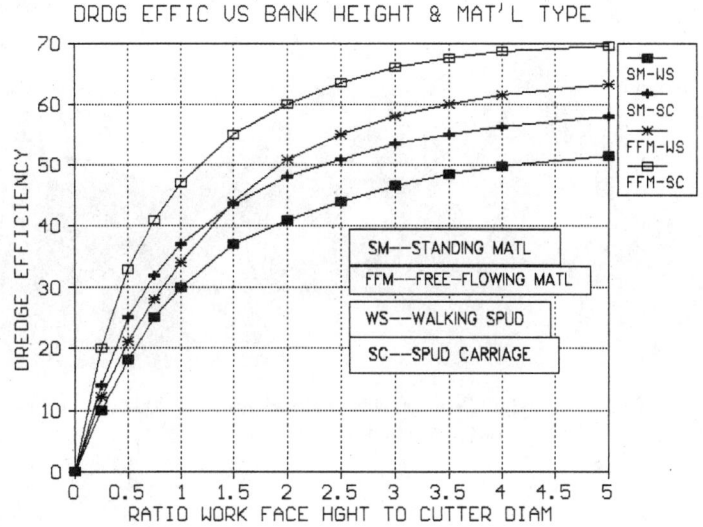

Fig. 3-2.

and that a walking spud dredge on standing material seldom achieves 50 percent D.E. Since the production rate of a dredge is directly proportional to D.E., it is obvious that failure to consider the major influences on D.E. can result in poor estimates. The estimator can select a preliminary D.E. from the plotted data of Fig. 3-2, the curves of which allow visual interpolation. D.E. is shown as a function of (1) work face height, (2) type of material, and (3) type of advance mechanism. Later a swing width correction is made if necessary to amend the chart value.

Fig. 3-2 is based on a dredge swing width of 200 feet (61 m); thus, a swing width other than 200 feet requires correction. Operators have noted that production rate increased with the width of the channel dug in a single swing. The wider channel allows more material to be dredged before the lost time of the dredge advance occurs; therefore the D.E. increases, increasing production rate. Some operators have added traveling spud carriages on spud barges secured to the stern of their dredge. This allows not only the faster action of the spud carriage advance but a wider potential swing as well, offering significant improvement in D.E.

The D.E. correction for swing width, either up or down, changes the ratio of productive time to non-productive time. Cal-

Fig. 3-3. "The 27" dredge *Illinois* with ladder pump, 1,200-horse-power cutter. Courtesy: Great Lakes Dredge & Dock Co.

culation of the change in D.E. can be simple, but an understanding of the swing cycle is necessary to obtain good results. A recording of the specific gravity of the slurry, as a dredge goes through its swing cycle, is a most informative document for an operator. It is not expensive to obtain during regular operation, and from this the various portions of the cycle can be identified and timed (spudding, production, etc.) and the dredge efficiency approximated. It shows that a wider swing increases production time, while spudding time remains the same. The recording can even identify the material as free-flowing or standing, an important consideration in D.E. determination.

While the D.E. chart has an empirical basis and has been used with considerable success, the operator may find that Fig. 3-2 and/or the swing width correction require adjustment for the idiosyncrasies of a given dredge. The operator should not hesitate to adjust for best results when proven data are available. A personal computer with appropriate software is highly recommended to determine D.E.

DREDGE LAW I REPHRASED

It should now be apparent that Dredge Law I can be rephrased by substituting the equality for average percent solids as defined in Dredge Law II.

DREDGE EFFICIENCY

Dredge Law I. Production varies as flow times average percent solids

Dredge Law II. Average percent solids equals maximum percent solids times dredge efficiency

Therefore Dredge Law I rephrased is:

Production varies as flow times maximum percent solids times dredge efficiency

SUMMARY

Production is a direct function of slurry flow rate, maximum percent solids, and D.E. Improvement in any of these elements without penalizing the others will have a salutary effect on production. There are various ways to improve D.E., and it is advisable for the operator to be familiar with them all.

Chapter 4

HYDRAULIC TRANSPORT FACTORS

DREDGE LAW III
MAXIMUM PERCENT SOLIDS VARIES WITH:
(A) VELOCITY IN THE SUCTION LINE;
(B) THE TYPE OF SOLIDS BEING DREDGED; AND
(C) INVERSELY AS THE SQUARE ROOT
OF THE DIAMETER OF THE SUCTION PIPE

The variances and their effect on dredging will be discussed below.

TURBULENCE REQUIREMENT FOR HYDRAULIC TRANSPORT

Hydraulic engineers have traditionally used the largest line economically justifiable to minimize turbulence and flow losses in a hydraulic system. In hydraulic transport, turbulence is mandatory to maintain the solids in suspension and to keep them moving at the approximate velocity of the fluid. If the turbulence is too low, the solids are not picked up at the suction inlet. If solids are picked up at a reasonable turbulence and subsequently the turbulence is reduced, the solids in the system tend to settle to the bottom of the pipe, creating a serious plugging potential. Maintenance of adequate velocity is a critical concern of the operator since velocity in the pipeline creates the necessary turbulence. The leverman controls velocity by controlling pump RPM, and velocity varies directly as pump RPM. The first variable, then, that determines the maximum percent solids that a dredge line can carry is the velocity in the suction pipe.

SUCTION VELOCITY

Frequently, it is advantageous to have a larger suction line than discharge line (see Chapter 8 for a further discussion of this topic) but the velocity at the suction inlet determines the capacity of the system. Obviously, all fluid and solids that enter the suction line must pass through the discharge line, and the discharge line if smaller will have a higher velocity and the solids will not settle out. If, however, the discharge line is larger than the suction, the velocity will be lower and the chance of settling out material in the line will be high. It is not good practice to have the discharge line larger than the suction, although a discharge line the same diameter as the suction has advantages when pumping moderate to long distances.

TURBULENCE REQUIREMENTS OF DIFFERENT MATERIALS

The second variable that determines the maximum percent solids that a dredge line can carry is the type of solids being conveyed. The settling velocity of denser, larger material is greater than that of lighter, smaller material; thus greater turbulence is required to keep the heavier material in suspension. The shape of the material can also affect its settling rate. It does not follow indefinitely that the larger a particle, the higher the required velocity. Very fine material has a large surface area per pound and a low settling rate as compared to larger material of the same density. However, for all practical purposes, the velocity that will successfully convey heavy gravel or cobbles will convey boulders. There is a system compensating effect that conveys boulders satisfactorily as long as they will pass through the pump impeller successfully. As an example, assume that gravel is being conveyed in a 12-inch pipe at 14 feet per second. If an 8-inch boulder enters the pipe, the area of the pipe would be reduced by 44 percent if the boulder were stationary. This would increase the velocity around the boulder in the line to 25 feet per second, providing more than adequate velocity to transport the boulder. This compensating velocity effect means that for all practical purposes, it is unnecessary to operate a dredge at velocities higher than those required for gravel unless a rare material is encountered with a specific gravity greater than 2.65, the normal rock and sand specific gravity.

Laboratory work has been done to ascertain the elements of a homogeneous material which affect its transport capabilities. From this work mathematical formulas have been derived which confirm that density, size, and shape are significant. However, in nature, a homogeneous dredge material is so rare as to be of little interest to the dredge operator. Empirical data on the actual material to be dredged is his practical answer, and there is data available for a broad spectrum of material. (See Limiting Velocity Chart, Fig. 4-1.)

Fig. 4-1 shows the velocities in feet per second required in 20-inch (508 mm) ID pipe to transport the indicated materials at the specified slurry specific gravities. The velocities required for different size pipes can be calculated easily by the technique shown in Equation 4-2. The chart in Fig. 4-1 is the only data required for a computer program to calculate velocities for all dredge sizes. With a PC, the software makes the necessary size conversions. The chart is based on granular 2.65 SG (2,650 g/l) solids, with 33.3 percent voids, and a 110 pcf (1,763 g/l) dry density.

VELOCITY REQUIREMENTS OF DIFFERENT PIPE SIZES

The last and most frequently overlooked factor affecting the percent solids a pipeline can carry is the diameter of the pipe. A brief analysis of the Fanning equation for friction head loss follows:

$$h = f \times \frac{L}{d} \times \frac{V^2}{2g} \qquad [\textit{Equation 4-1}]$$

$$\text{Or: } V^2 = \frac{2g \times d \times h}{f \times L}$$

Therefore: $V \sim (d)^{0.5}$

The above is true if we assume that the unit head loss (i.e., turbulence) for hydraulic transport of a given material at a given concentration is constant regardless of line size (a reasonable approximation). Then, the required velocity will vary as the square root of the line diameter. This means that a higher velocity is required in a larger line to transport the same percent solids of a given material the same distance.

The velocity required in any size line can be calculated if the satisfactory velocity is known for any other line size. For example,

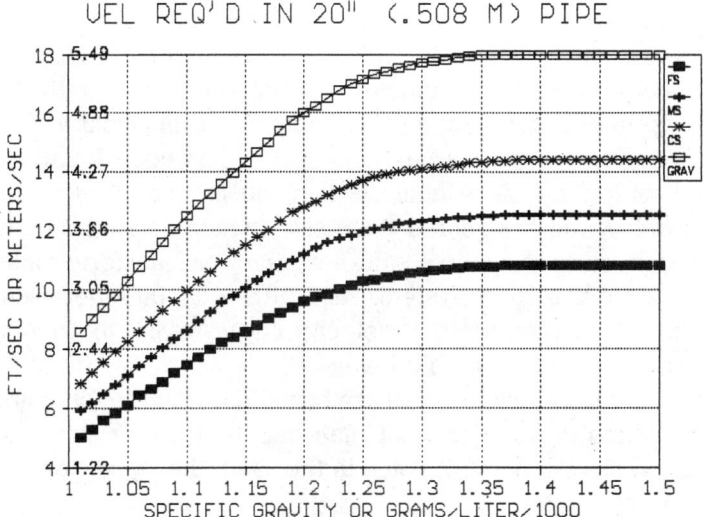

Fig. 4-1. Velocity limiting curve.

if 10 ft/sec velocity is satisfactory in a 10-inch inside diameter pipe, the velocity required in a 20-inch pipe would be:

$$\frac{V_2}{V_1} = \left(\frac{d_2}{d_1}\right)^{0.5}$$

[Equation 4-2]

$$V_2 = \left(\frac{20}{10}\right)^{0.5} \times 10 = 14.14 \text{ ft/sec}$$

This can be confirmed by checking Fig. 4-1, the limiting velocity curve for coarse sand. This curve is based upon collected empirical data for 8-inch inside diameter pipe and extrapolated for other sizes of pipe. The curves are invaluable to the operator in determining the necessary velocity for a cost-efficient operation.

It is important to note it is the suction velocity that determines the percent solids picked up. Obviously, all material that passes the suction passes the discharge at the same flow rate, GPM, but not necessarily at the same velocity. Frequently, the suction pipe is larger than the discharge, so that the same GPM provides different velocities in the different size lines. Velocity is the key to hydraulic transport. If a satisfactory velocity can be maintained in a dredge line at all times, the dredgeman's problems are greatly simplified.

The limiting velocity curves of Fig. 4-1 define fine sand as having a median grain size of 0.1 millimeter; medium sand 0.32 millimeter; coarse sand 1.0 millimeter; and gravel as 10.0 millimeters. The dredgeman practically never encounters a material without a broad spectrum of grain sizes. However, it has been found that a material with a d_{50} sieve analysis (i.e., 50 percent above and 50 percent below the designated grain size) which corresponds to either fine, medium, or coarse sand or gravel will perform similarly in a hydraulic transport system. Also, if the d_{50} falls between the grain sizes for which limiting velocity curves exist, extrapolation of velocities gives reasonable results.

The limiting velocity curves are based on a 2.65 specific gravity material which is true for most materials that the dredgeman encounters. With 33.3 percent voids in the sand, the weight of the dry sand is only:

$$.667 \times 2.65 \times 62.4 = 110 \text{ lbs/cu ft or } 49.89 \text{ kg}$$

With zero voids, the weight would be:

$$1.00 \times 2.65 \times 62.4 = 165 \text{ lbs/cu ft or } 74.83 \text{ kg}$$

but of course, if a dredgeman encounters 165-pound material (solid rock), he must disintegrate it in some fashion in order to hydraulically transport it. Therefore, the voids are an inevitable part of his operation.

The specific gravity coordinate of the curves is based upon the use of 1.0 SG liquid which introduces a negligible error in the event seawater is the conveying medium. Tests and actual dredge experience have shown that the dredgeman should always strive to achieve a 1.5 SG slurry, because here the most cost- and power-effective hydraulic transport is achieved. This corresponds with 45.4 percent in situ volume. Above this point, the ratio of water to solids is less conducive to efficient transport, and eventually when the only water is that which fills the voids in the sand, there is no water available for hydraulic transport, and the resistance to flow approaches infinity.

SOIL CLASSIFICATION

The Unified Soil Classification System (USCS) is generally used in the U.S., but it was not developed for the dredging industry and has some shortcomings. It specifies all material passing the 200 mesh screen as silt; yet we know sand exists below 200 mesh. USCS also specifies coarse sand as passing the #4 sieve (slightly under one fifth of an inch or 5 mm), which many observers would consider pea gravel. Recognizing the limitations of the USCS, the author utilizes a convenient system for dredge calculations that allows a logical progression from the grain size of one category to the next by a multiple of 10. PIANC, the respected European organization involved with ports, navigation, and dredging, has independently developed a system that has a striking similarity to the author's system. There are other systems as well that the dredgeman may encounter.

Fig. 4-2 delineates the size range for each named soil of the three systems above. While the soil names are the same, the ranges differ somewhat; however, the PIANC and Turner systems have midpoints that coincide for Fine, Medium, and Coarse Sands, as well as Gravel. The USCS system ranges vary from the other two, but there is some similarity. The advantage of the Turner system is that the ranges are the same size on the logarithmic scale (but with Medium Sand superimposed as the midrange of Fine and Coarse Sand). The ranges are convenient and easy to remember,

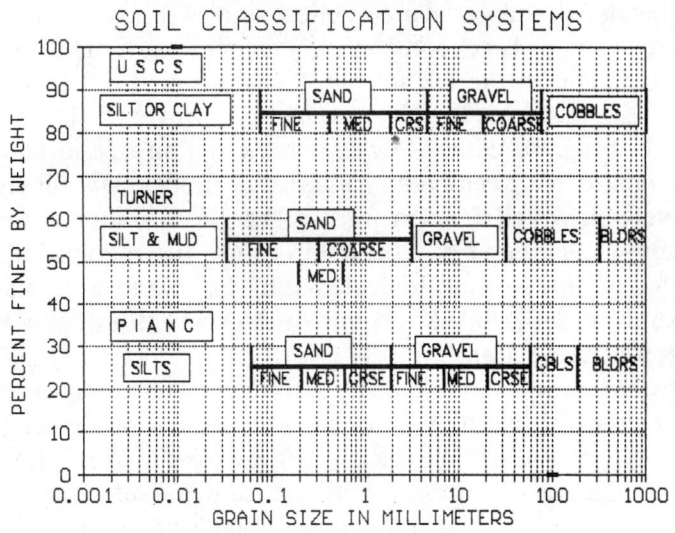

Fig. 4-2

with each midpoint coinciding with the beginning of each log section, .01, .1, 1, 10, and 100.

It should be emphasized that any of the soil systems can be used by the dredgeman if proper care is taken. By always referring to the sieve analysis, the confusion caused by the various systems' names for differing material ranges is resolved by ignoring the names and using the median grain size. The d_{50} is the determining hydraulic transport factor, not the name.

Nevertheless, names for dredged materials are useful and convenient. They have been used traditionally by geotechs and dredgemen, and would be difficult to eliminate. A program to calculate dredge capacity should encompass the common materials Silt, Fine Sand, Medium Sand, Coarse Sand, and Gravel; but it is important that the d_{50} and/or size range for each material be clearly defined.

It is essential that the dredgeman know which soil system is utilized in the project specifications to avoid name confusion. He also needs to understand the rudiments of geotechnology where it affects the performance of his cutter or slurry system. See Chapter 13, "Cutters".

Although the dredgeman does not need all the expertise of the geotech, there are aspects where his perspective must override that of the geotech. For example, when deriving a d_{50} from the sieve analysis of a material for production calculations, the dredgeman must consider the material's condition as it reaches the hydraulic transport system. The geotechnical engineer may disintegrate a hard-pan sample in the lab for sieve analysis and obtain a d_{50} of .316 mm, the midpoint for Medium Sand; however, if the material fails to disintegrate completely under the action of the cutter (a likely scenario with cemented sands), the dredgeman should use higher velocities and friction losses than those required for Medium Sand. It is a good rule of thumb to use the rheological characteristics of coarse sand for calculating cemented Fine and Medium Sands and Clays that do not disintegrate thoroughly under the action of the cutter.

There is a story told about the self-taught dredgeman who had great success with his 10-inch dredge on coarse sand, while pumping at a 10 foot per second velocity. He calculated that if he had a 20-inch dredge, he could run it with the same personnel, and since it would have 4 times the flow of the 10-inch dredge *at the same velocity* (area varies as diameter squared), he expected 4 times the

Fig. 4-3. Dredge *Wheeler*, the Great Flagship of the U.S. Army Corps' of Engineers, with three dragarms (two overside and one in centerwell). Hopper capacity 8,000 cu yd. Courtesy: U.S. Army Corps of Engineers.

production and much improved profits. So he invested all of his accumulated savings in a new 20-inch dredge.

By referring to the coarse sand limiting velocity curve, it is apparent that while 10 feet per second provided about 45 percent solids (in situ slurry volume) in his 10-inch dredge, his 20-inch dredge, at 10 feet per second provided an in situ slurry volume of only 8 percent solids. His production (according to Dredge Law I) dropped as follows:

$$\text{Production} \sim [4 \times (\text{flow})] \times [8/45 \text{ percent solids}] = .711$$

Or only 71.1 percent of the production of the 10-inch dredge.

Therefore, our dredgeman was financially and professionally embarrassed when his new, larger dredge produced at a lower rate than his old, smaller unit. He had failed to recognize that as pipelines increase in size, they require higher velocities to achieve the turbulence required to carry the same percent solids.

SUMMARY

The percent solids that can be carried in a pipeline is a function of the slurry velocity, the nature of the solids, and the size of the pipeline. Turbulence is required to keep the solids in suspension and flowing. Larger lines require a higher velocity to achieve the equivalent turbulence of a smaller line at a lower velocity. The dredgeman does not require all the soils expertise of the geotech, but he must be able to interpret the geotech data. Recognizing the various soils and their effect on the performance of his dredge is essential.

Chapter 5

MAXIMUM DREDGE PRODUCTION

DREDGE LAW IV
THE MAXIMUM OUTPUT OF A DREDGE VARIES AS THE AREA OF THE SUCTION PIPE

The maximum output of a dredge occurs on short discharge lines where there are no pump head or velocity limitations. We shall see that the limitation is barometric.

BAROMETRIC HEAD INDUCES FLOW

Fig. 5-1 represents a tank filled with water to a level of h feet. Water exits through the nozzle at velocity V, which is a function of the water head available to create V. This relationship is expressed by the velocity head term developed in Chapter 1 as follows:

$$h = \frac{V^2}{2g} \qquad [Equation\ 1\text{-}1]$$

By eliminating the constants, we can say:

$$h \sim V^2$$

By rearranging and taking the square root of both terms, we see that:

$$V \sim \sqrt{h}$$

There is obviously barometric pressure equal to 34 feet of water on the surface of the water in the tank. This does not add to the head available to generate velocity through the nozzle because the same barometric pressure is external to the tank resisting the flow of water through the nozzle. However, if we could discharge the nozzle into an infinitely large tank and evacuate the tank to an

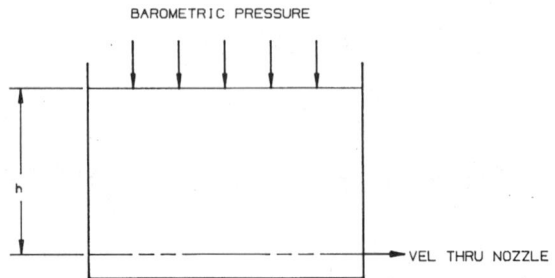

Fig. 5-1. Velocity through nozzle.

absolute head of 6 feet of water, i.e., a vacuum of 28 feet, then we would add 28 feet of water to the head h, increasing V as follows:

$$V \sim \sqrt{h + 28}$$

Now, if we assume the water level in the supply tank drops to the nozzle level, the h becomes zero and:

$$V \sim \sqrt{28}$$

By substituting a dredge pump for the impossible concept of an infinitely large tank, we now have a practical analog of the hydraulic system of a dredge. See Fig. 5-2. Most dredge pumps are mounted with their centerline at or near the waterline, and a good pump can pull a maximum vacuum of about 25 inches of mercury or 28 feet of water without fear of cavitation. Since there is no static water head on the suction side of the pump, *the only force available to induce flow to the pump is barometric pressure*, or the 28 feet of it that the pump can utilize.

DREDGE FLOW VARIES WITH SUCTION LINE VELOCITY AND AREA

The dredge pump is a device that evacuates its casing; it cannot mechanically reach down the suction pipe to pick up the slurry. But since nature abhors a vacuum, its ubiquitous barometric pressure rushes in to fill the vacuum. To reach the evacuated pump casing, however, barometric pressure must pass through the suction pipe and through the excavated soil placed at the suction mouth, thus creating the velocities needed for hydraulic transport.

MAXIMUM DREDGE PRODUCTION   43

D.L.IV   MAX.OR SHORT LINE PRODUCTION ~ A_s

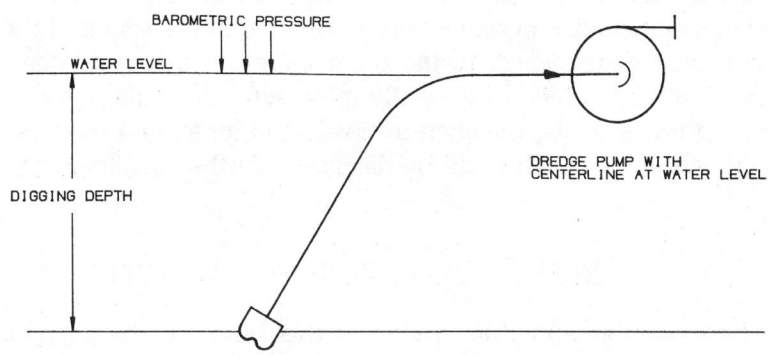

SINCE h IS BAROMETRIC PRESSURE AND A CONSTANT, MAX.V IS CONSTANT
SINCE FLOW = V_s X A_s THEN FLOW ~ A_s
SINCE PRODUCTION VARIES AS FLOW (DREDGE LAW I), IT ALSO VARIES AS A_s

Fig. 5-2. Dredge suction system.

The dredge pump then picks up the slurry provided by this barometric pump, and generates its own pressure from the impeller velocity as explained in Chapter 1. The slurry then passes through the discharge line to the disposal area; thus, the dredge performs its hydraulic function.

The suction line is the only access to the suction side of the pump; and the only force to induce velocity through the cross-sectional area (A_s) of the suction pipe is barometric pressure. Therefore:

   Flow = $V_s \times A_s$

Since we know that the maximum suction velocity, V_s, is a function of the square root of 28 at sea level, then V_s is a constant. It follows then that:

   Flow maximum ~ A_s
   Since production ~ flow × average percent solids
                                                    [*Dredge Law I*]
   Then maximum production ~ A_s [*Dredge Law IV*]

In reality, the dredge pump is a booster pump to nature's barometric pump (the weight of the atmosphere that forces water to the pump), and as every experienced dredgeman knows, when the dredge pump is run faster than the barometric pump can supply it, cavitation results. It has been the author's experience that the most significant dredge design errors have been in the sizing of the suction pipe with respect to the other elements of the dredge. Also, the discharge pipe is frequently mis-sized. As a rule, when a line-sizing error exists, the suction line is too large, and the discharge line is too small. This will be discussed further in Chapter 8.

EFFECT OF ALTITUDE ON VELOCITY

The examples and calculations in this book are based upon the approximate barometric pressure at sea level of 30 inches mercury, 34 feet water, or 14.7 psig. Since this pressure is induced by the head of the air column above the earth's surface, measured at sea level, it follows that if we measure the barometric pressure at increasing elevations, the head diminishes. This is analogous to the pressure reduction on an object which is rising from the depths of a body of water. Since both water and air are fluids, the effect of lessening the fluid column on head is identical.

The table showing the effect of altitude on the density and barometric pressure of air is shown in Fig. 5-3. Note that at sea level (zero altitude) the barometric pressure is 29.92 inches mercury; at 1,000 feet altitude 28.85 inches; at 3,000 feet, 26.81 inches; while at Denver, Colorado (using 5,200 feet altitude for the mile high city), the barometric pressure is only 24.71 inches mercury or 82.6 percent of the pressure at sea level.

Does this reduction in force affect the capacity of a dredge operating in Denver? Very much so. The only force available to push the slurry to the evacuated dredge pump is barometric pressure, and that force is reduced to $.826 \times 34 = 28$ feet of water.

Since the imperfect dredge pump is unable to utilize the last 6 feet of the barometric pressure, 6 feet must be deducted from the 28 feet available, leaving only 22 feet to overcome the Denver suction losses. Since suction velocity V_s varies as the square root of available barometric head, $V_s = \sqrt{h_B}$

$$V_{max} \sim \left(\frac{22}{28}\right)^{0.5} = 0.886$$

MAXIMUM DREDGE PRODUCTION 45

ALTITUDE: DENSITY TABLE FOR AIR
Standard Air at 0 Altitude and 29.92" Bar = 1.00
Altitudes in Feet   Constant Temperature = 70° F

Alt	Rel Den	Bar	Alt	Rel Den	Bar	Alt	Rel Den	Bar
0	1.00	29.92	2000	0.930	27.82	5000	0.832	24.89
100	0.996	29.81	2100	0.926	27.72	5200	0.825	24.71
200	0.993	29.70	2200	0.923	27.62	5400	0.819	24.52
300	0.989	29.60	2300	0.920	27.52	5600	0.813	24.34
400	0.985	29.49	2400	0.916	27.41	5800	0.807	24.16
500	0.981	29.38	2500	0.913	27.31	6000	0.799	23.98
600	0.978	29.28	2600	0.909	27.21	6500	0.786	23.53
700	0.975	29.17	2700	0.906	27.11	7000	0.774	23.09
800	0.971	29.06	2800	0.903	27.01	7500	0.758	22.65
900	0.967	28.96	2900	0.900	26.91	8000	0.739	22.12
1000	0.964	28.85	3000	0.896	26.81	8500	0.728	21.80
1100	0.960	28.75	3200	0.889	26.61	9000	0.715	21.38
1200	0.957	28.65	3400	0.883	26.42	9500	0.701	20.98
1300	0.954	28.54	3600	0.877	26.23	10000	0.687	20.57
1400	0.951	28.44	3800	8.870	26.03	15000	0.564	16.88
1500	0.947	28.33	4000	0.864	25.84	20000	0.458	13.70
1600	0.944	28.23	4200	0.858	25.65	25000	0.371	11.10
1700	0.940	28.13	4400	0.851	25.46	30000	0.297	8.88
1800	0.936	28.02	4600	0.845	25.27	35000	0.235	7.03
1900	0.933	27.92	4800	0.839	25.08	40000	0.185	5.54

Fig. 5-3. Altitude effect on air density and barometric pressure.

This is a reduction of 11.4 percent in maximum velocity, but is only the tip of the iceberg with respect to dredge capacity. The discussion in Chapter 6 will disclose the need to redistribute the available suction head over the various losses in the suction line so as to optimize production. Depending upon digging depth, material being pumped, size of dredge, etc., production could be lowered much more than 11.4 percent as a result of the loss in barometric head pressure caused by the Denver altitude.

EFFECT OF ALTITUDE AND TEMPERATURE ON HORSEPOWER

There is yet another potential loss in dredge capability because of altitude. Note that the density of air in Denver is only 0.832 of that at sea level. Diesel engines require a great deal of air to supply the oxygen necessary to efficiently burn the oil injected into the cylinders. If the air is rarefied, i.e., reduced in density, the engine HP must be down-rated.

A further down-rating of the engine HP may occur if the temperature of the intake air is too high. Note Fig. 5-4 in which a loss of air density is shown as the temperature increases. Air for inter-

TEMPERATURE: DENSITY TABLE FOR AIR*

Temp in deg Fahr Standard Air = 70° F = 1.00

Temp	Dens	Temp	Dens	Temp	Dens	Temp	Dens	Temp	Dens
-10	1.178	60	1.019	100	.946	200	.803	400	.616
-5	1.165	62	1.015	105	.938	210	.791	425	.599
0	1.152	64	1.011	110	.930	220	.779	450	.582
5	1.140	66	1.008	115	.922	230	.768	475	.567
10	1.128	68	1.004	120	.914	240	.757	500	.552
15	1.116	70	1.000	125	.906	250	.747	525	.538
20	1.104	72	.996	130	.898	260	.736	550	.528
25	1.093	74	.992	135	.891	270	.726	575	.512
30	1.082	76	.989	140	.883	280	.716	600	.500
35	1.071	78	.985	145	.876	290	.707	625	.488
40	1.060	80	.982	150	.869	300	.697	650	.477
42	1.056	82	.978	155	.862	310	.688	675	.467
44	1.052	84	.974	160	.855	320	.680	700	.457
46	1.047	86	.971	165	.848	330	.671	725	.447
48	1.043	88	.967	170	.841	340	.662	750	.438
50	1.039	90	.964	175	.835	350	.654	775	.429
52	1.035	92	.960	180	.828	360	.646	800	.421
54	1.031	94	.957	185	.822	370	.638	825	.412
56	1.027	96	.953	190	.815	380	.631	850	.404
58	1.023	98	.950	195	.809	390	.624	875	.397

*Density in this sense is relative density or more properly specific gravity.

Fig. 5-4. Temperature effect on air density.

Fig. 5-5. Split hull trailing suction hopper dredge *Eagle 1*. Courtesy: C. F. Bean Corporation.

nal combustion engines should normally be taken from outside the engine room, since it is possible to encounter engine room temperatures as high as 130°F, which result in air density losses exceeding 10 percent.

SUMMARY

The only force available to a conventional dredge to push slurry to the dredge pump is barometric pressure. When barometric pressure is utilized fully, the maximum velocity occurs in the suction line and maximum capacity is achieved as a function of the area of the suction pipe. Altitude above sea level reduces barometric pressure and air density for fuel combustion, and therefore can cause an appreciable reduction in dredge productivity.

Artist's version of the Suez Canal's dredge, *Mashour*, 30,785 HP, the world's most powerful cutterhead dredge, scheduled for delivery in 1996. Courtesy: IHC.

Chapter 6

THE SUCTION LINE AND DIGGING DEPTH

DREDGE LAW V
THE OPTIMUM SUCTION LINE VELOCITY VARIES WITH THE DIGGING DEPTH

Years ago, the author visited a large dredge which was working on the harbor of Marseilles, France. The cutter was set at a digging depth of 7 meters. Upon completing a cut, the operator of the dredge lowered the cutter to 10 meters and immediately adjusted his pump speed upward. When asked why he had increased pump speed, he responded, "Everyone knows it takes more horsepower to pick material up from 10 meters than it does from 7 meters. I just gave it more horsepower."

And indeed he had, for he increased substantially the water he was pumping. But, he had also reduced his solids payload significantly and was getting nothing in return for his increased fuel consumption except decreased production, increased wear, and additional water problems in the disposal area. This chapter explains why the operator's action was incorrect, and what he should have done.

ANALYSIS OF SUCTION LINE LOSSES

A fundamental grasp of the functions and problems of the suction line is the key to understanding the hydraulic dredge. Most problems with the pumping system originate in the suction line; its equilibrium is delicately balanced, easily disturbed, and sorely restricted by the barometric limitation. This chapter examines the losses in the suction line, the understanding of which is essential to the proper design and operation of a hydraulic dredge.

In a hydraulic dredge without a ladder pump, the only force to overcome the suction line losses is the barometric pressure afforded by the column of air above the earth's surface. This is the equivalent of 34 feet (10.35 meters) of water head on the surface of the water being dredged. A conventional dredge pump can utilize 28 feet (8.54 meters) of that amount to feed the pump during normal operation.

This is referred to as "vacuum" in the suction line, i.e., 28 feet less than barometric pressure, or more accurately as 6 feet of absolute pressure. If the operator exceeds the 28-foot vacuum, he runs the risk of cavitation (see Chapter 11). The task of the operator under most conditions is to utilize the full 28 feet to maximize productivity, but to avoid cavitation. To complicate his problem further, the operator must distribute the 28 feet over the various losses in the most effective manner to optimize production.

There are four ever-present losses in the suction line of an operating dredge, and a fifth if the dredge pump is above the water line: (1) velocity head—the head necessary to accelerate the slurry from standstill to the required transport velocity; (2) entrance loss—the head required to force the slurry into the suction mouth; (3) friction loss—the head to overcome frictional resistance to flow in the pipe; (4) specific gravity head—the head to lift the solids in the slurry from the bottom of the channel to the centerline of the pump; and (5) lift head—the head to lift the slurry from the waterline to the pump centerline (a loss when the pump is above the waterline, and a gain when below).

VELOCITY HEAD

The velocity head term is common to all hydraulic calculations, appearing in most equations. Its formula is $H_v = (V^2/2g) \times SG$. Since water has an SG of 1.0, the SG term can be ignored when dealing with water alone; however, when dealing with slurries, it is essential. Logically, more force is required to accelerate a slurry of 1.5 SG to a given velocity than for 1.0 SG water—50 percent more as demonstrated by the equation above.

By ignoring the constants of 2g and SG in the H_v equation, it can be seen that H_v varies with the square of velocity. Conversely, in this completely reversible relationship, the velocity varies with the square root of the available head; therefore, in order to double the velocity, it is necessary to quadruple the head. Obviously, this is not a straight line relationship, but it is not all bad; if we double the tip speed of a pump impeller, we get four times the head, a most fortuitous circumstance for the users of centrifugal pumps.

With respect to the suction line, it is important to recognize that as velocity is increased, more of the barometric pressure must be utilized to provide the necessary velocity head. Fortunately, the required head only increases as the square root of the velocity; however, since only 28 feet of barometric pressure is available, any increase in velocity head demands a corresponding decrease elsewhere. The operator must utilize the valuable 28 feet in the most efficient manner, because excessive velocity can reduce percent solids and productivity.

ENTRANCE LOSS

Entrance loss has traditionally gotten little industry attention in proportion to its importance. There is much more consciousness of friction in the suction line, but seldom does friction rise to the level of entrance loss. Entrance loss deserves to be better understood.

The formula for entrance loss is $H_e = K \times (V^2/2g) \times SG$. This equation is identical to H_v except for the entrance coefficient, K. Cameron Hydraulic Data shows K varying between .04 and .5, but this is for Newtonian fluids and can mislead the unwary dredgeman. The slurry attempting to crowd its way into the suction mouth is a far cry from a Newtonian fluid, and requires a much higher coefficient. In the event of a cave-in, when the suction mouth can be covered by many feet or meters of material, the coefficient can approach infinity, causing instant cavitation. For a suction line to pick up the desired slurry of 1.5 SG, the suction mouth must be essentially covered by a shallow depth of granular solids, thus assuring the high solids content required for economic operation. If a portion of the mouth is not covered, the water takes the path of least resistance, entering the pipe without its solids load. The solids-covered suction mouth requires a higher entrance coefficient. An entrance coefficient of 1.0 has successfully emulated the dredging operation. Conceivably, it may be high for light-maintenance material and low for clays, but it has worked well in general applications. This makes the entrance loss equivalent to that of the velocity head for slurry, as it, too, must be multiplied by the SG of the slurry.

FRICTION LOSS

The Darcy-Weisbach equation for friction loss is $H_f = F \times (L/D) \times (V^2/2g) \times SG$, where F is the friction factor, L is the line length, and D is the inside diameter of the pipe in feet. This equation is not as convenient as the Hazen-Williams equation on slurries (see Chapter 10), but is helpful in showing the relationship of the H_v term to suction line analysis. By calculating the friction loss for water with Darcy-Weisbach and multiplying by SG, a reasonable figure is achieved.

SPECIFIC GRAVITY HEAD

The equation for specific gravity head is $H_{SG} = DD \times [SG_s - SG_w]$. Note the equation calculates solids lift only, since the water SG is deducted from the slurry SG. Water seeks its own level. With the pump centerline at water level, no lift is required for water; however, the solids require a lift from the bottom to the pump centerline (the digging depth).

It should be emphasized that the suction line losses are calculated with the maximum SG, not the average. The dredge pump does not cavitate at the average SG, but when the maximum SG demands the highest vacuum. Average SG is used successfully in calculating the losses in the long discharge line, because the long line contains peaks and valleys of SG, resulting in automatic averaging; however, the short suction line requires the use of maximum SG.

SUCTION LIFT

Most dredge pumps are mounted at the waterline, and a suction lift is not required. When the dredge pump is above or below the

THE SUCTION LINE AND DIGGING DEPTH

waterline, a lift (positive or negative) is required as shown by the equation $H_L = L \times SG_s$. Note that unlike H_{SG}, where solids only had to be raised to the pump, here both water and solids must be lifted. It is a serious error to build a dredge with its pump above the waterline, unless it has a ladder pump to provide the lift. To deplete the limited barometric head for a suction lift reduces the dredge capacity significantly and unnecessarily. A 3-foot lift would require 4.5 feet of barometric head at 1.5 SG. This would reduce the available head by $4.5/28 \times 100 = 16$ percent. This would reduce velocity and capacity by almost 30 percent.

An increase in capacity can be achieved by mounting the pump below the waterline within the dredge hull. This increase is modest because of the physical limitation on lowering the pump in the shallow conventional hull. Most designers conclude that the increase is not worth the hazard and/or the inconvenience of being unable to remove the stone box cover for fear of sinking the dredge. Unfortunately, this has occurred.

OPTIMIZING SUCTION VELOCITY

Since the dredge pump makes only 28 feet of barometric head available, it must be distributed among the various suction losses. For each calculation, the sum of the losses is equated to 28 feet, and then the maximum (optimum) velocity is calculated. Obviously, as depth is increased, velocity is decreased. There is an optimum suction velocity for every depth, the determination of which is the first step in calculating dredge rate. Fig. 6-1 is an approxi-

Parameter	Water	Slurry		
	30-ft depth 1.0 Sg	30-ft depth 1.5 SG	50-ft depth 1.36 SG	50-ft depth 1.24 SG
H_v	3.3	5	4	6
H_e	1.7	5	4	6
H_f	2.0	3	2	4
H_{SG}	0	15	18	12
H_L	0	0	0	0
Total	7.0	28	28	28

FIG. 6-1. Typical suction losses for 24 inch dredge.

Fig. 6-2. Dredging International's hopper dredge *Pearl River*, with 17,000 m³ hopper capacity, the largest trailer afloat in 1995. Courtesy: IHC.

mation of the suction losses for the conditions shown. Note that at the 30-foot depth, the dredge achieves the desirable 1.5 SG, utilizing the entire 28 feet of barometric pressure. At the 50-foot depth, velocity has been sacrificed for H_{SG}, but even then, an SG of only 1.36 is achieved. This means that when the depth is increased from 30 to 50 feet, the velocity and SG are both reduced so that the capacity of the dredge is decreased exponentially. If the operator makes the mistake of increasing the suction velocity at the expense of H_{SG} when depth is increased (Fig. 6-1, last column), the capacity is further reduced. More water is pumped, but the solids, the dredge's payload, are reduced.

The 7-foot total under the first column in Fig. 6-1 (water) represents the "water vacuum" for the 30-foot depth before solids enter the suction line. The 28 feet under the other columns represents maximum practicable vacuum before cavitation.

THE SUCTION LINE AND DIGGING DEPTH

SUMMARY

The natural tendency of the operator to increase suction velocity as digging depth increases results in increased costs and lower production. There is an optimum or correct velocity for every depth. This optimum velocity results in the distribution of the 28 feet of barometric pressure over the several suction losses so as to maximize production rate. An understanding of the suction line and its losses is essential to the efficient operation of the hydraulic dredge. A computer program that calculates the optimum suction velocity is a valuable aid to the operator. Once this velocity is established, other data helpful to the leverman fall into place, e.g., head, specific gravity of slurry, friction loss per 100 feet of line, cubic yard per hour, maximum line length, and HP. Knowing a dredge's capability can prove highly motivational to its operators.

Chapter 7

HORSEPOWER VS. LINE LENGTH

DREDGE LAW VI

**LINE LENGTH VARIES
AS PUMP HORSEPOWER**

During an extended trip through South America, the author was invited aboard a large, well-maintained dredge working in a major harbor. Obviously, the operators took great pride in the impressive dredge, which they had purchased in the United States. It was equipped with conventional dredge instruments, vacuum gauge, pressure gauge, RPM indicator, but they did not have a production meter. A powerful diesel engine drove the dredge pump, and it was tended lovingly by the watch engineer.

During operation, it became apparent that the leverman did not control the pump speed. Investigation disclosed that the watch engineer established the pump engine speed (regardless of operating conditions) by observing the exhaust manifold temperature and adjusting engine speed to maintain the maximum allowable manifold temperature, in order to provide maximum horsepower output. These conscientious operators reasoned that if they were maximizing the pump horsepower, they were maximizing production. Since maximum centrifugal pump horsepower coincides with maximum velocity which is frequently inimical to high production, we know from Dredge Laws IV and V that this reasoning is fallacious. Not only were the operators needlessly burning fuel and wearing out components with excessive velocity, but they were reducing production by severely limiting the solids percentage of the slurry.

Dredge pump horsepower *is* a significant element and influences productivity under various circumstances. However, there is

HORSEPOWER VS. LINE LENGTH

no consistent correlation between maximum horsepower and maximum production. In order to examine the significance of pump horsepower to dredge production, we should start with the pump horsepower equation which follows:

$$HP = \frac{GPM \times 8.34 \text{ lbs/gal} \times \text{specific gravity} \times h}{33,000 \text{ ft lbs/min} \times \text{pump efficiency}} \qquad [Equation\ 7\text{-}1]$$

GPM, American gallons per minute, times 8.34 represents pounds per minute of water. When multiplied by the specific gravity of the slurry, pounds per minute of slurry is obtained. When further multiplied by head, we obtain foot pounds per minute, a true horsepower expression. Since there are 33,000 foot pounds per minute in one horsepower, dividing by that value determines the horsepower demand of the fluid. The pump is not 100 percent efficient, so the equation must also incorporate the pump efficiency in the denominator to obtain the drive horsepower required at the input end of the pump shaft.

HORSEPOWER VS. GPM, SG, AND h

By eliminating the numerical constants in the horsepower equation, and ignoring pump efficiency which changes with gallons per minute, we have a very convenient and useful form of the horsepower equation.

$$HP \sim GPM \times SG \times h \qquad [Equation\ 7\text{-}2]$$

This says that horsepower varies directly with gallons per minute, specific gravity, and head required. Any change that the operator makes in any of these elements has a direct and predictable effect on horsepower.

By going a step further and eliminating all elements in Equation 7-1 that we know should remain constant, we can relate HP to any remaining variables. The reader will recall Dredge Law V which states the velocity in the suction line should be a constant at a given digging depth. Obviously, we know the depth at which the dredge is digging so that the velocity (and therefore the GPM term in the equation) is constant. The specific gravity will be held at a constant maximum for production reasons, and will result in a constant average in the discharge line as a function of dredge efficiency. Pump efficiency is a constant at a constant GPM as shown on the pump characteristic curve.

This leaves the only variable as the head, h, and head is directly proportional to line length. For example, if the head loss is five feet per 100 feet of line, and there are 4,000 feet of line, including suction, the loss is:

$$5 \times \frac{4,000}{100} = 200 \text{ ft}$$

This loss is within the capability of a single dredge pump, most of which would have a head-generating capability of between 200 and 260 feet. If the line length were only 3,000 feet, the head requirement would be 150 feet, and the pump could be slowed down and the horsepower correspondingly reduced. On the other hand, if the job requirement were 8,000 feet, the head requirement would be 400 feet. This would require a booster pump with the same horsepower as the dredge pump.

It should be apparent that with gallons per minute and specific gravity constant, the horsepower will vary directly as head, which varies directly with line length. Therefore, the line length against which a dredge can pump is a function of the horsepower available on the pumping system, Dredge Law VI.

The explanation above assumes that there is a reasonable compatibility between the pump and its drive, i.e., that the drive has the horsepower required to turn the impeller at the gear ratio provided, and that the impeller can absorb the horsepower and transmit it to the fluid. For further discussion of pumps and drivers, see Chapter 13. For a graphical picture of the effect of the pump horsepower, see the production charts in Chapter 8.

PIPELINE SIZE VS. FRICTION

The flow through a pipeline is the product of the velocity times the area of the pipe.

$$Q = V \times A \qquad [Equation\ 7\text{-}3]$$

Where Q = cu ft/sec, V = ft/sec and A = sq ft

HORSEPOWER VS. LINE LENGTH

To convert to the more common GPM:

$$GPM = V \times A \times 60 \text{ sec/min} \times 7.48 \text{ gal/ft}^3$$
$$GPM = 448.8 \times V \times A \qquad [Equation\ 7\text{-}4]$$

To convert A to the more commonly used inside diameter of the pipe in inches:

$$GPM = 448.8 \times V \times \frac{\pi d^2}{4 \times 144}$$
$$GPM = 2.448 \times V \times d^2 \qquad [Equation\ 7\text{-}5]$$

$$\text{Or } V = \frac{GPM}{2.448 \times d^2} \qquad [Equation\ 7\text{-}6]$$

Where A = Cross sectional area of pipe in square feet
V = Velocity in ft/sec
d = Inside diameter of pipe in inches
GPM = Gallons per minute

h_F VS. GPM AND PIPE DIAMETER

The Fanning or Darcy-Weisbach equation for Friction Head is:

$$h_F = f \times \frac{L}{d} \times \frac{V^2}{2g} \qquad [Equation\ 4\text{-}1]$$

By eliminating constants, we can express the convenient relationship between friction head, velocity, and inside pipe diameter.

$$h_F \sim \frac{V^2}{d} \qquad [Equation\ 7\text{-}7]$$

By substituting the value of V in Equation 7-6 and dropping the constant, we obtain:

$$h_F \sim (GPM/d^2)^2 \div d \sim \frac{GPM^2}{d^5} \qquad [Equation\ 7\text{-}8]$$

From this relationship we can see that if GPM is held constant and the pipe diameter is doubled, the unit friction head become 1/32 of its former value. (Actually, the empirical Hazen-Williams equation indicates friction varies inversely with pipe diameter to the

4.8655 power rather than 5.0.)

$$(GPM_2/GPM_1)^2 \div (d_2/d_1)^5 = (1/1)^2 \div (2/1)^5 = \frac{1}{32}$$

Of course, the converse is true. If the pipe diameter is halved, the friction increases.

$$(1/1)^2 \div (1/2)^5 = \frac{1}{.03125} = 32$$

Seldom does the occasion arise where the dredgeman makes a practical decision to halve or double his line size. However, he is constantly making decisions between, e.g., 12-inch and 14-inch lines, and this is a common ratio of discharge to suction line on dredges shown as follows:

$$(1/1)^2 \div (12/14)^5 = \frac{1}{.4627} = 2.16$$

This indicates that the unit friction head of a 12-inch inside diameter pipeline is more than twice that of a 14-inch inside diameter pipeline for a given GPM. This can be confirmed by checking against any friction table which uses the Fanning equation, e.g., the Cameron Hydraulic Data. The same calculation for 20- and 24-inch pipe indicates a 2.49 greater loss for the smaller pipe.

EFFECT OF SUCTION SIZE ON PUMPING DISTANCE

As the sixth Dredge Law states, the dredge discharge line length capability is a function of the HP of the pumping system; there is, however, another factor that affects line length: suction line size.

Perhaps the most common error in dredge design occurs in sizing the suction line with respect to the discharge line. If the suction line is too large, low velocity causes low solids pick-up and low production. On the other hand, if the discharge line is too small, a high price is paid in friction, wear, and fuel.

See Fig. 7-1 for a computer-generated graph representing production for a 20-inch dredge, with three different suction sizes. The dredge is identical in all respects other than suction size. A case can be made in favor of either suction line size as a function of digging depth and line length. The following table of examples is derived from Fig. 7-1.

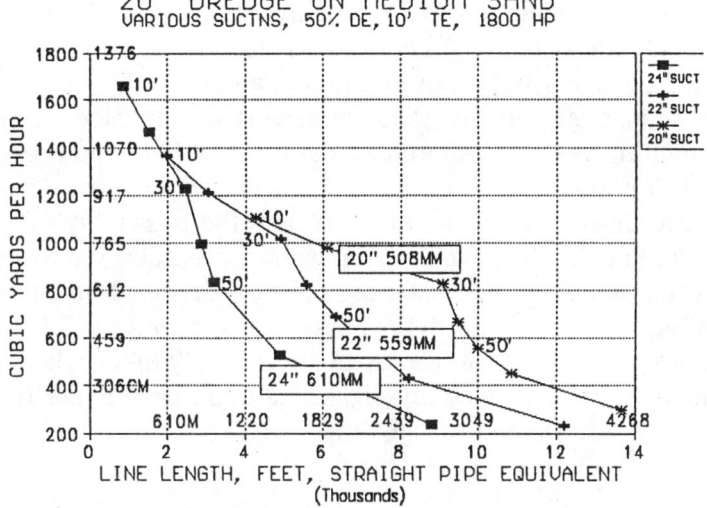

Fig. 7-1. 20-inch dredge with various suctions.

Suction Diameter	Digging Depth	Line Length	Production Rate
24	30	2,500	1,200*
22	30	2,500	1,010
20	30	2,500	825
24	30	5,000	510
22	30	5,000	980*
20	30	5,000	825
24	30	9,000	220
22	30	9,000	380
20	30	9,000	825*

At a 30-foot depth, the 24-inch suction has a clear production advantage (*) at the 2,500-foot line length; the 22-inch is superior at 5,000; and the 20-inch at 9,000 feet. Under the specified conditions, the 20-inch suction never causes a discharge line length limitation, but it is barometrically limited to a maximum rate of 825 cu yd/hr. The 24-inch suction becomes velocity limited at 2,500 feet, the 22-inch at 5,000 feet, and the 20-inch at 9000.

It is apparent that the optimum suction line size is a function of project conditions. If different depths or line lengths are used than those above, the production advantage will shift. Thus, there is an optimum ratio of suction to discharge line size *as a function of the project conditions*.

Without the computer to develop production charts such as Fig. 7-1, the dredge operator functions at a disadvantage. See Chapter 8 for a further discussion of production charts.

The most efficient hydraulic transport occurs near the minimum limiting velocity required by the material being pumped. See Fig. 4-1. It is specious reasoning to "play it safe" by using a smaller discharge line to avoid "settling out" and the possibility of "plugging" the line. This possibility is actually increased by the smaller line when pumping long distances because the head of the pump is limited, and the unit friction head is increased by 2.16 times when using a 12-inch line versus 14-inch, or 2.49 times when using a 20-inch versus a 24-inch line. By using Equation 7-8, the friction loss can be ascertained for any line sizes.

HORSEPOWER VS. LINE SIZE (HORSEPOWER COEFFICIENT)

The amount of horsepower made available to the dredge pump has varied significantly from builder to builder throughout the years. One builder, e.g., recommended the use of a 3,600-horsepower diesel engine for its standard 30-inch dredge pump. Others have felt that a 5,000-horsepower drive was better. Some dredge pump builders have gone as high as 6,000 and 7,000 horsepower. The horsepower variation has not been restricted to 30-inch units, but wide variations have been apparent through the entire range of dredge design from 10 inches up through 36 inches.

Dredge Laws III and VI provide insight as to the requirement of horsepower when a dredge line size is established. Dredge Law III states that to carry the optimum percent solids of a given material in a given size line, the velocity in the line must be such as to provide a relatively high level of turbulence. Then, if the line size is increased, *flow* will necessarily increase as the square of the ratio of the line sizes at the same velocity; however, the *velocity* must increase as the square root of the ratio of the new line to the old to maintain transport turbulence. As a result, horsepower must increase still further.

Dredge Law VI states that the distance a dredge can pump its production is directly proportional to the horsepower on the dredge pump. It also develops from the pump horsepower equation

HORSEPOWER VS. LINE LENGTH

that:

$$HP \sim GPM \times SG \times h \qquad [Equation\ 7\text{-}2]$$

For the purpose of developing a horsepower constant to guide in powering any size pump, we shall assume the following regarding the above relationship:

1. The specific gravity of slurry will be the same regardless of line size. The operator should attempt to optimize his solids content at all times.
2. The product line length requirements will not vary regardless of whether a small or a large dredge is used. The material must be moved from its in situ location to the disposal area.
3. The optimum friction loss per 100 feet of line for a given concentration of a given material is relatively constant, regardless of line size. This was previously established in connection with Dredge Law III.

Acceptance of the above assumptions leaves horsepower varying solely as GPM (Equation 7-2). Since GPM for a given velocity varies as the area of the pipe and as the square of the diameter, it is apparent that horsepower will also vary as the square of the pipe diameter against which it is pumping. However, Dredge Law III points out that as a pipeline increases in diameter, the turbulence in the pipe at a given velocity decreases. Therefore, in order to retain the turbulence required to convey the same percent solids, the velocity must increase to the .5 power as the pipe size increases. Thus, horsepower varies as:

$$d^2 \times d^{.5} = d^{2.5}$$

This states that the horsepower required to pump a given concentration a constant distance varies as line diameter to the exponential of 2.5. Therefore, we can establish a convenient horsepower coefficient, C_{HP}, by dividing the horsepower available to the dredge pump by the discharge line diameter to the 2.5 exponential.

$$C_{HP} = \frac{HP}{d^{2.5}} \qquad [Equation\ 7\text{-}9]$$

$$\text{Or } HP = C_{HP} \times d^{2.5} \qquad [Equation\ 7\text{-}10]$$

$$\text{Or } d = \left[\frac{HP}{C_{HP}}\right]^{0.4} \qquad [Equation\ 7\text{-}11]$$

An 18-inch dredge pump which has 1,125 horsepower would have a horsepower coefficient of:

$$C_{HP} = \frac{HP}{(d)^{2.5}} = \frac{1,125}{(18)^{2.5}} = 0.82$$

Likewise, a 24-inch dredge pump with 2,250 horsepower would have:

$$C_{HP} = \frac{2,250}{(24)^{2.5}} = 0.80$$

Fig. 7-2 shows twelve other examples of cutterhead dredge pumps and drives. All of the units (plus the two above) have been successful except for numbers 3 and 7. Both of these units had excessive wear and cavitation problems due, at least partially, to the fact that their discharge lines were too small for effective utilization of the horsepower available. Their horsepower coefficients confirm this.

No.	Size	HP	C_{HP}
1	30	3,600	.73
2	30	5,000	1.01
3	27	5,200	1.37
4	27	2,875	.76
5	24	2,875	1.02
6	22	2,250	.99
7	20	2,250	1.26
8	20	1,700	.95
9	16	970	.95
10	16	850	.83
11	14	725	.99
12	12	480	.96

Fig. 7-2. Pump horsepower coefficients for existing cutterhead dredges.

None of the above pumps was unsatisfactory because of *low* horsepower coefficient. Problems of cost, size, and wear arise with high horsepower coefficient, but low horsepower coefficient results only in limited pumping distance. Therefore, if a project has a short line length, a low horsepower coefficient could prove satisfactory. For example, a 24-inch hopper dredge with 1,000-horsepower pump is a viable design, and has a horsepower coefficient of only 0.35. The lowest horsepower coefficient for a cutterhead dredge known to the author was 0.51 where the special design dredge was used for creating canals in a delta and was pumping approximately 1,000 feet. Such a design would not prove successful as a general contractor's dredge unless a booster pump were available.

It appears that a good cutterhead dredge pump value of horsepower coefficient is 1.0. A satisfactory range would be 0.5 to 1.2 with the understanding that the low coefficient has limited pumping distance and the high coefficient has a tendency toward high wear. By adding booster pumps, the pumping distance can be extended as far as desired without excessive wear. It is questionable practice to have a single pump coefficient over 1.2 on abrasive materials.

RECOMMENDED PUMP HORSEPOWER

Fig. 7-3 is a tabulation of recommended horsepower ranges for dredge pumps sized from 6 through 42 inches. These values have been rounded off, and the designer should deviate from these figures to a reasonable degree in order to utilize standard, available drives.

Dredge size in inches	Horsepower at approximate horsepower coefficient of		
	0.7	0.85	1.0
6	60	75	100
8	125	150	200
10	225	275	325
12	350	425	500
14	500	625	750
16	700	875	1025
18	950	1175	1400
20	1250	1525	1800
22	1600	1950	2250
24	2000	2400	2800
27	2650	3200	3800
30	3450	4200	5000
33	4400	5300	6300
36	5500	6600	7800
42	8000	9700	11500

Fig. 7-3. Recommended pump horsepower vs. line size.

The author readily acknowledges that many "successful" dredges have applied pump horsepower in excess of the above. Such dredges are candidates for larger pipe, which would result in greater capacity, longer pumping distances, less wear, and greater

Fig. 7-4. Concept of frameless dredge, ladder and spuds powered by hydraulic cylinders. Courtesy: Ellicott Machine Corporation.

profits. See Fig. 8-3 for the effect on production of increasing a 20-inch line to 24 inches with a 2,250 horsepower drive.

SUMMARY

It is difficult to overemphasize the importance of the relationship of suction line size to discharge line size; friction head loss to line size; and pump head and horsepower to discharge line size. These relationships will determine the viability of the dredge as an economic unit.

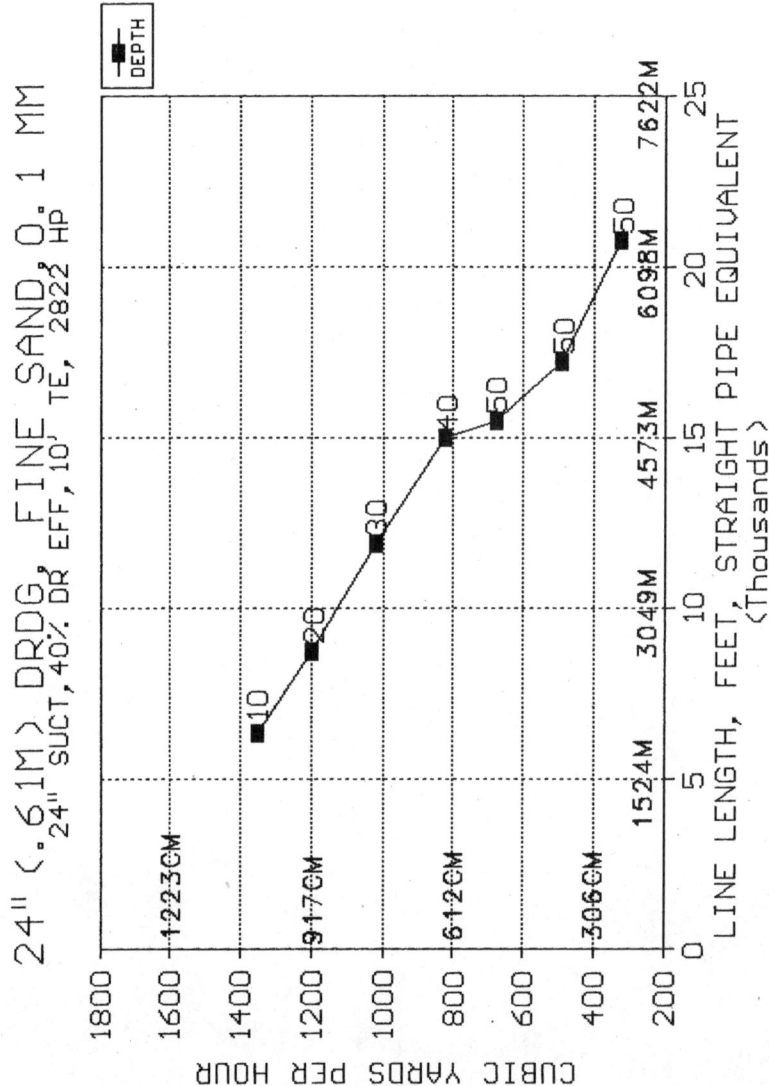

Fig. 8-0. Production chart—24-inch dredge, 24-inch suction.

Chapter 8

PRODUCTION CHARTS

DREDGE LAW VII
**PRODUCTION IS LIMITED BY:
(A) SUCTION CONDITIONS (BAROMETRIC HEAD);
(B) PUMP HP (DISCHARGE HEAD REQUIREMENT);
AND (C) SLURRY VELOCITY (TRANSPORT CAPABILITY)**

The seventh Dredge Law is a summary statement of the first six laws. The production limitations expressed in Dredge Law VII are clearly evident in the shape of the production charts shown in Figs. 8-0 through 8-7. Each figure has been calculated and produced by a proprietary PC program developed by the author. The subject of the figures is a 24-inch × 24-inch (610 mm) dredge with a dredge efficiency of 40 percent and conventional pump characteristics of HP, eye speed, and tip speed. A HP coefficient of 1.0 is used (2,822 HP); a maximum eye speed of 42 ft/sec (12.8 m/s); and a tip speed of 113 ft/sec (34.45 m/s).

The production charts plot rate (cu yd/hr or cm/hr) against equivalent straight line discharge length (feet or meters). Various digging depths are shown to demonstrate the reduced rate at deeper depths. On Fig. 8-0, the points at 10, 20, 30, 40, and 50 foot depths represent the meeting of the pump's barometric limitation (ordinate) and its head limitation (abscissa). The point at 40 feet is where the limiting velocity is encountered, i.e., where the maximum SG can no longer be carried in the suction line because of inadequate velocity. Production falls rapidly after this point. The last two 50-foot points on the chart are not barometrically limited, but use arbitrarily reduced SGs to demonstrate the dredge's long-line transport capabilities.

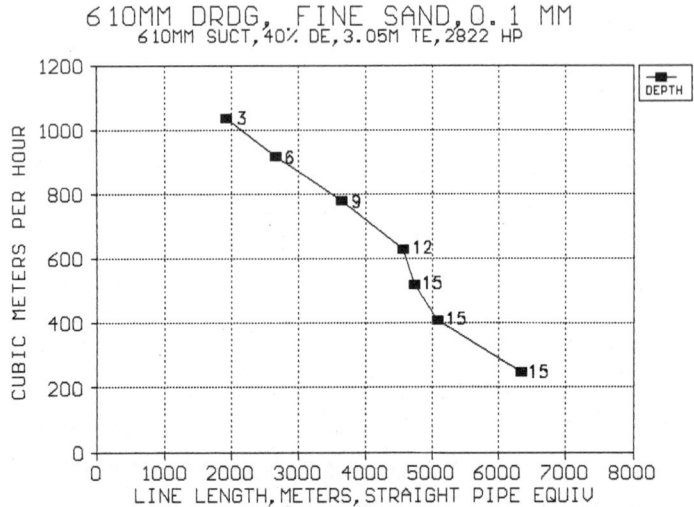

Fig. 8-1. Production chart—610-mm dredge, 610-mm suction.

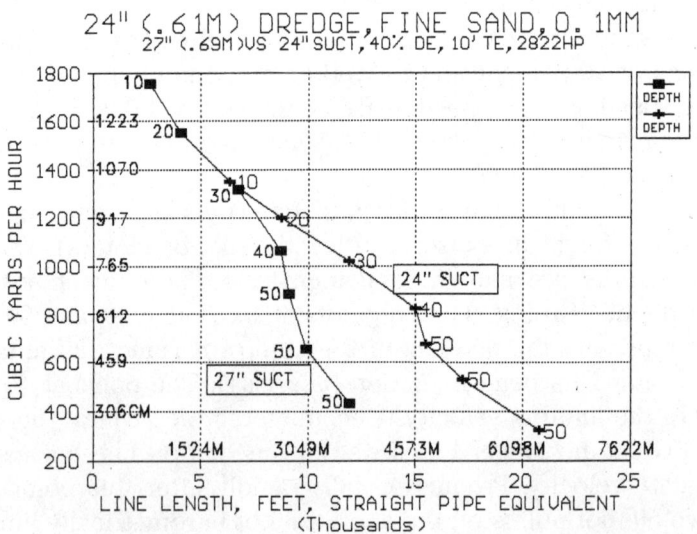

Fig. 8-2. Production chart—24-inch dredge. 24-inch vs. 27-inch suction.

PRODUCTION CHARTS

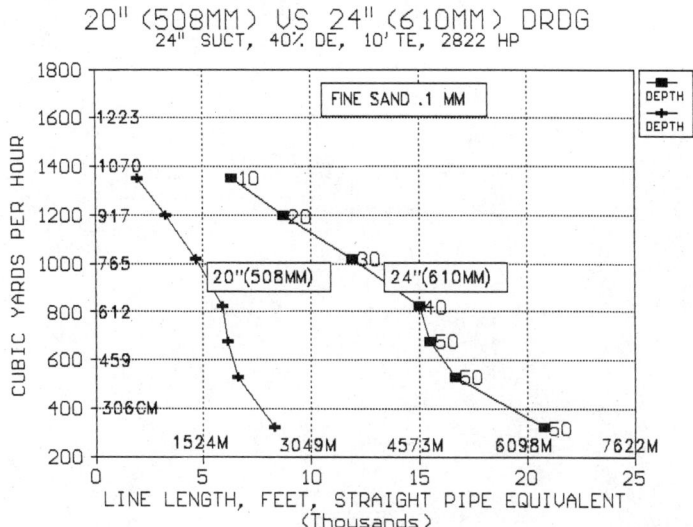

Fig. 8-3. Production chart: 24-inch suction, 20-inch vs. 24-inch discharge.

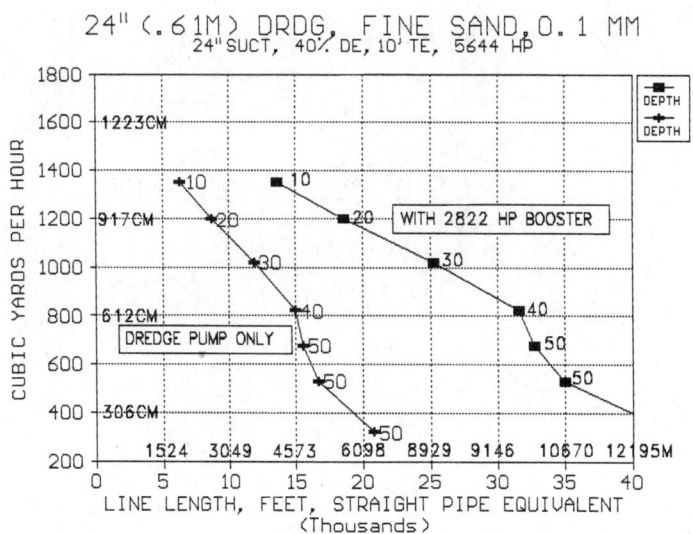

Fig. 8-4. Production chart: 24-inch dredge, with and without booster pump.

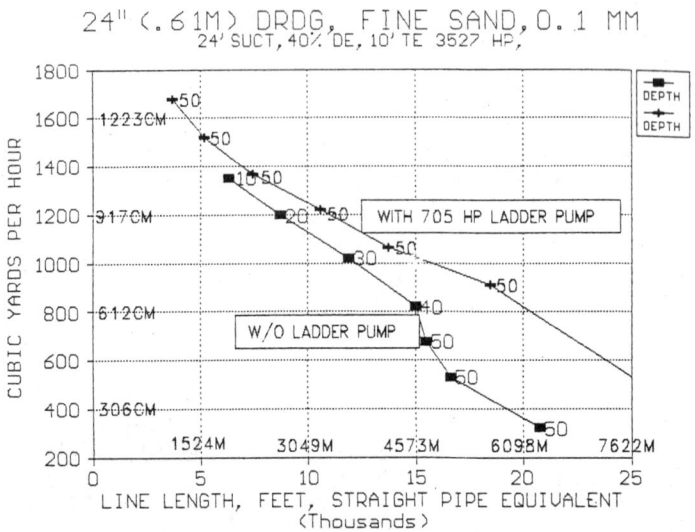

Fig. 8-5. Production chart: 24-inch dredge, with and without ladder pump.

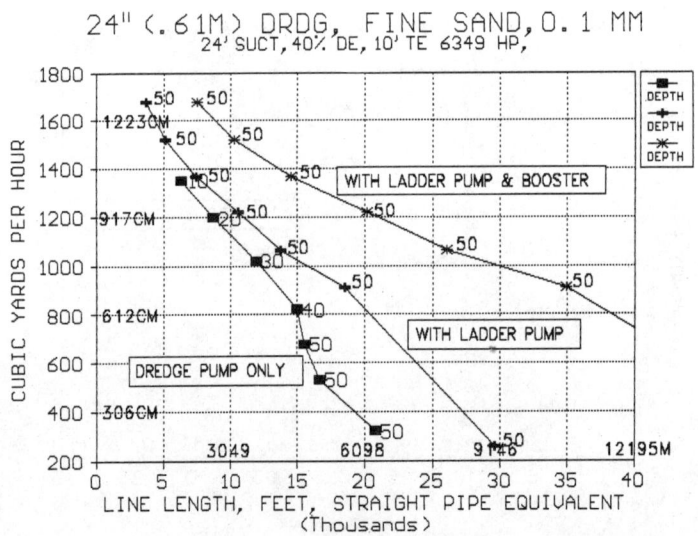

Fig. 8-6. Production chart: 24-inch dredge, with ladder pump and booster.

PRODUCTION CHARTS 73

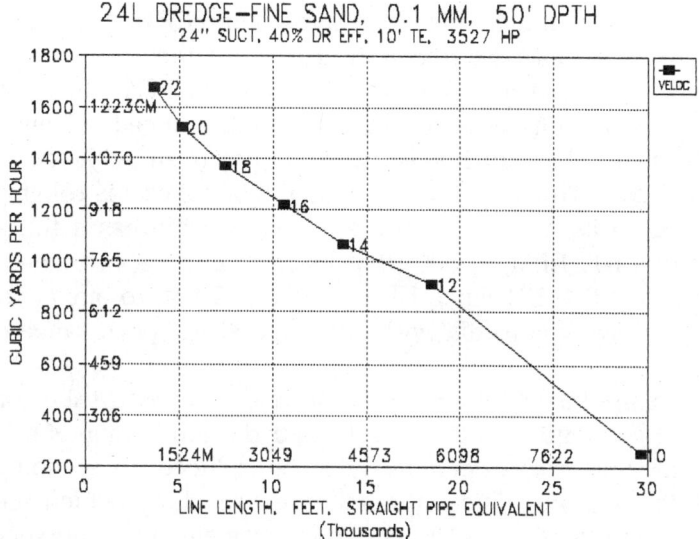

Fig. 8-7. Production chart: 24L dredge including dredge and ladder pumps.

Each chart in this chapter demonstrates an important dredging principle. The serious student of hydraulic dredging is encouraged to study these charts and explanations carefully. Understanding the charts and the underlying dredge laws constitutes a powerful tool for the dredgeman.

Dredge Law V (Chapter 6) indicates that when digging depth increases for a non-ladder pump dredge, production decreases. Fig. 8-0 confirms this, showing the maximum rate allowed by the barometric pressure at each designated depth (the point labels). The rate is plotted against the maximum line length at each point, with length limited by HP and/or head of the pump. Logically, the lower rates can be pumped further.

Fig. 8-0 shows production rate at a 10-foot digging depth to be about 1,350 cu yd/hr at any distance up to 6,300 feet. To pump less than the maximum chart distance, the operator would reduce pump RPM to reduce head, but he should still pump the same optimum GPM of slurry, achieving the same 1,350 cu yd/hr. If the drive is restricted by its speed range, it is possible a minimum length of discharge line may have to be utilized to provide resistance to establish the GPM of the pump at a reasonable velocity for optimum operation.

As the discharge line lengthens, the pump RPM should increase to provide the head and horsepower to pump the slurry the greater distance. At 4,000 feet, the torque limitation of the drive is encountered. The drive is normally not at full speed at this point, which means that full horsepower has not been achieved on the diesel engine. The horsepower capability of a diesel is roughly proportional to its speed, i.e., if a 2,000 HP engine has a full speed rating of 1,000 RPM, at 800 RPM its HP will be approximately 80 percent or 1,600 HP. Since HP = torque × RPM, torque capability remains roughly constant within the operating speed range of the engine.

At this point of torque limitation, the pump can transport the 1,350 cu yd/hr no further. Therefore, if the job demands a longer line, the cubic yards per hour must be reduced. In so doing, the lower production reduces the solids in the slurry which reduces resistance in the line, lowering two major elements of horsepower, specific gravity, and head (Equation 7-2).

As the line length increases, the production rate continues to fall in order to keep the foot-pounds requirement of the hydraulic transport system compatible with the capability of the pump and drive. As the line increases in length, the pump eventually reaches full speed and full horsepower (not apparent on the chart). At this point the operator has no control left other than to reduce the pickup of solids, lowering the specific gravity of the slurry. As the discharge line continues to lengthen, the velocity and production continue to fall (Dredge Law I), since RPM can no longer be increased. At about 15,000 feet the velocity drops to the critical range as indicated in the limiting velocity charts (explained in Dredge Law III, Chapter 4). From this distance and beyond, the production of the dredge drops so drastically that it raises the question of economic justification of continuing without the addition of a booster pump.

At each digging depth, the maximum output is represented by the horizontal line at the designated depth. For example, at 30 feet digging depth, the output is about 1,000 cu yd/hr. This production can be maintained with proper operating techniques out to a distance of 12,000 feet where the torque limitation is encountered. The chart then follows the same pattern as though the digging depth were 10 feet because the dredge is no longer limited by barometric pressure, but by the drive torque capability which at some line length is common to all digging depths.

At 50 feet digging depth the maximum output is less than 700 cu yd/hr, one half of the maximum output at 10 feet digging depth. This suggests a useful rule of thumb for the dredgeman: *The maximum output of a conventional cutterhead dredge at 50 feet digging depth is one-half of its output at 10 feet digging depth.* The 50 feet digging depth horizontal production line intersects the outer envelope of the chart at 15,500 feet (Fig. 8-0), where the chart is common to all digging depths. Fig. 8-1 shows the same data as Fig. 8-0, but is expressed solely in the metric system for clarity.

SUCTION LINE SIZE

Dredge size is commonly expressed as the discharge line size. In Fig. 8-0, the 24-inch dredge also has a suction line of 24 inches. In each of Figs. 8-2 through 8-6, the production chart from Fig. 8-0 appears for comparison purposes. Each new chart shows the effect of some design change on the slurry system, and is explained in the text.

Fig. 8-2 shows the effect of increasing the suction line from 24 inches (610 mm) to 27 inches (686 mm). The short-line production of the 27-inch suction is significantly increased over that of the 24-inch. This is in accordance with Dredge Law IV (Chapter 5), which states the maximum production of a dredge varies with the area of the suction pipe. An obvious production advantage exists for the 27-inch suction on "short" lines; however, at about 8,000 feet, the velocity limitation of the 27-inch suction line is reached and production drops rapidly. Actually, there is an advantage for the 27-inch suction out to about 9,000 feet if the digging depth is 40 feet. Note that the maximum output of the 24-inch suction is slightly over 800 cu yd/hr at a 40-foot depth, which corresponds to the 27-inch rate at 9,000 feet.

Velocity limitation is not reached on the 24-inch suction until 15,000 feet, so a distinct production advantage is demonstrated for the smaller suction diameter on "long" lines. The production rate of the 24-inch suction doubles that of the 27-inch at 15,000 feet. Also, if the digging depth were 10 feet, there would be an advantage for the 24-inch suction at all lengths beyond 7,000 feet. *The most advantageous size suction is a function of job conditions.* To say "use a large suction on short lines, and a small suction on long lines" may have merit, but is too simplistic. The dredgeman needs the data shown on the chart, including depth, to make the right decision. For example, the 27-inch suction provides 880 cu yd/hr (673 m^3/hr) at a 50-foot (15 m) digging depth and 9,000-foot (2,744 m) line length. The 24-inch suction provides only 680 cu yd/hr (520 m^3/hr) under those conditions; however, if the line length were 15,500 feet (4,725 m), the 24-inch suction would give the same 680 cu yd/hr while the 27-inch would give only about 200 cu yd/hr. The data supplied by the computer is essential to the operator's ability to make the correct decisions for varying job conditions.

If a dredge is to be used on projects of unpredictable line length, it is feasible to use a suction somewhat larger than the discharge line, but rarely if ever should it be more than 12.5 percent larger (the 27/24 ratio). This provides a suction area 26.6 percent greater, so the long line production rate will be compromised as Fig. 8-2 shows. The author has seen dredges with disparately large suctions that severely hampered output, e.g., a 34/27 ratio where the suction was 58.6 percent greater than the discharge. Here, in order to induce the desirable 15 SG slurry on coarse sand in the suction at 18.7 ft/sec, an exorbitant discharge velocity of 29.6 ft/sec would occur, resulting in a very short line length capability and high wear. A larger discharge should be used under such circumstances, unless the dredge pump is underpowered, in which case the suction should be reduced.

DISCHARGE LINE SIZE

Fig. 8-3 shows the effect of replacing the 24-inch discharge line with a 20-inch size on the subject dredge while retaining the 24-inch suction. This 24-inch/20-inch configuration has been widely used in practice, but it should be noted that there is a difference of 20 percent in diameter and 44 percent in area.

Since the suction line is 24 inches in both cases, the maximum or short line production is 1,350 cu yd/hr (Dredge Law IV). However, the torque limitation of the drive on the 20-inch discharge line is reached at about 2,000 feet, compared to more than 6,000

PRODUCTION CHARTS

feet for the 24-inch line. The resistance of the smaller line versus the larger is doubled at the same GPM.

The author recalls two cases where different line sizes were used with this basic dredge, one successfully and one not so successfully. A European company wanted a dredge to pump a soft, loamy material a few thousand feet. A suction and discharge line of 27 inches was applied to the dredge (no change in the 24-inch pump). The material dug easily, dispersed well, and partially because of the lighter organics, conveyed readily. The job was eminently successful, and the dredge owner completed the job in about two-thirds of the scheduled time.

The second case was the application by an American owner on sand, again on a few thousand feet of line. Here, however, the owner had thousands of feet of 20-inch pipe on hand which he used over the protests of the dredge builder. The result was, as shown in Fig. 8-3, a low production beyond 6,000 feet, plus high fuel costs and high wear on the pump and pipeline.

The first owner held the opinion that his dredge pump was the finest unit in the industry. The second owner held a much less flattering view of his unit. Of course, the pumps were identical, making the point that even the best of equipment can fail to perform satisfactorily if improperly applied.

Numerous dredges exist today with discharge lines 44 percent smaller in area than their suction lines. *There is little or no merit to this relationship;* it reduces production rate and line length, and increases HP and wear rate. In most such cases (small dredges are less affected), owners would be well advised to change to a larger discharge line, but never larger than the suction.

It is feasible to replace worn pipe sections with a new larger line, operating with two sizes of pipe in a common discharge line. The sizes should not be intermixed, i.e., all the smaller size pipe should be contiguous. This avoids the high friction losses incurred by multiple size transitions.

BOOSTER PUMP EFFECT

Fig. 8-4 shows the effect of adding a 2,822-horsepower booster pump in the discharge line of the dredge. Dredge Law VI states that the line length against which a dredge can pump is proportional to pump horsepower. Note that the original dredge pump, also 2,822 horsepower, had the line length capability of slightly more than 6,000 feet. With the booster pump, the distance is increased to 13,000 feet, somewhat more than twice as far. The explanation for the extra few hundred feet is that the resistance of the suction line was overcome by the dredge pump in both cases, leaving the booster some extra horsepower for discharge line length.

Note the effectiveness of the booster pump in increasing production on long lines. At 20,000 feet, the dredge pump alone would produce less than 400 cu yd/hr. With the booster added, the production is just under 1,200 cu yd/hr at a digging depth of 20 feet (6 m).

LADDER PUMP EFFECT

Fig. 8-5 shows the effect of adding a submerged pump to the otherwise identical dredge. The HP of the ladder pump is 705, 25 percent of the dredge pump's default value. Whereas previously the maximum dredge output was 680 cu yd/hr at the 50-foot digging depth, it now shows just under 1,700 cu yd/hr. However, the dredge pump HP has not changed, and line length limitations still apply. The small amount of ladder pump HP not utilized for the suction does increase system head, allowing a slightly longer discharge line.

The ladder pump should be designed to create enough head (perhaps 60–90 feet as a function of depth) to overcome all suction losses and provide a modest positive pressure at the dredge pump. This breaks the barometric bottleneck and prevents cavitation. A larger ladder pump with more HP and head could be used, but the effect would only be longer discharge line capability. Longer lines can normally be achieved more economically with surface-mounted equipment; however, on short line operations such as sand and gravel, it can be more economical to use a ladder pump only, i.e., no hull pump. The second pump is avoided by increasing the ladder pump HP and head sufficiently to overcome the resistance of the short discharge line.

PRODUCTION CHARTS

The suction line size of a ladder pump dredge should be the same as the dredge discharge line. Some dredges have used a larger suction, perhaps as a holdover from the conventional dredge. The reason for the larger suction on the conventional dredge is to mitigate the effects of the limited barometric pressure. Since the ladder pump eliminates the barometric limitation, the larger line unnecessarily reduces the pickup velocity, lowering the potential SG and production rate. It also causes SG fluctuations, complicating rate calculations.

In Fig. 8-5, the reader should not be confused by the close proximity of the ladder pump curve to that of the dredge pump alone. There is a large advantage to the ladder pump dredge, because all rates are achievable at a 50-foot digging depth. The conventional dredge will produce less than 700 cu yd/hr at a 50-foot depth, whereas the ladder pump dredge produces up to 1,700 cu yd/hr on short lines. The advantage of the ladder pump diminishes (but never disappears) when the dredge is digging at shallow depths. See Chapter 14 for further discussion.

Fig. 8-6 shows a 2,822 HP booster added to the 24-inch ladder pump dredge of Fig. 8-5. This provides a significant increase in operating range for the higher production of the non-barometrically limited dredge.

Fig. 8-7 shows the 24-inch "L" dredge chart on fine sand. The point values are the same as the middle curve on Fig. 8-6, but the point labels are slurry velocity, not depth as used on the "D" dredges. Depth is an insignificant factor for the "L" dredge as the ladder pump breaks the barometric bottleneck.

Fig. 8-8 shows the 24-inch x 24-inch "D" dredge on four common materials. Having the materials on a common chart highlights the differences in production rate and line length, attributable to the materials alone. In planning the project, failure to distinguish between fine sand and coarse sand would inject a significant error. At 10,000 feet, the chart shows the dredge has over twice the capacity on fine sand as on coarse sand. A computer program to highlight these facts in graphical form is invaluable to the operator.

Fig. 8-9 shows the 24-inch "L" dredge on the same materials as Fig. 8-8. The "L" dredge has great advantages over the "D" dredge. Its first and operating costs are somewhat higher, but its increased capacity at depths greater than 25–30 feet is such as to justify the extra costs easily.

THEORIES OF DREDGING

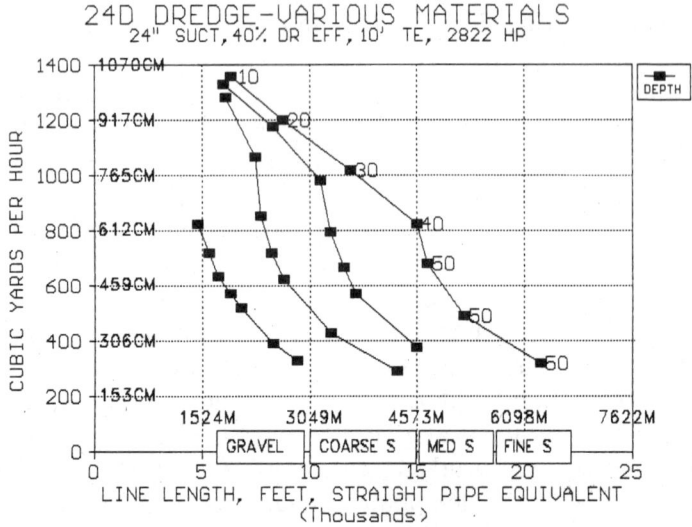

Fig. 8-8. Production chart: 24D dredge for various materials.

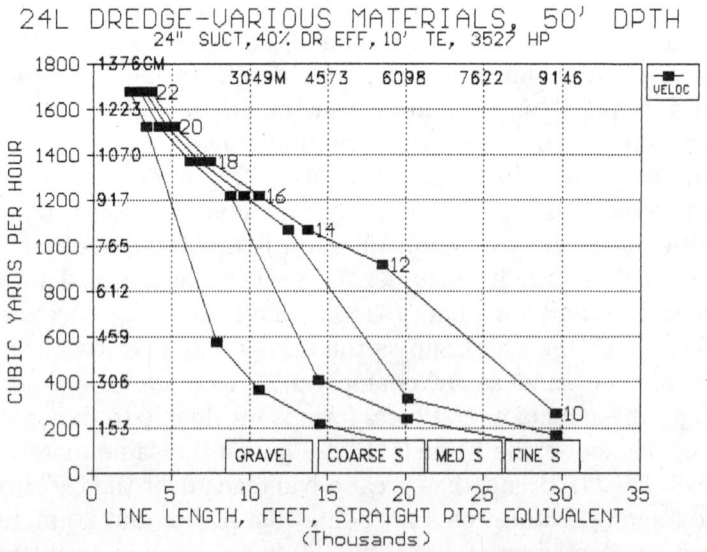

Fig. 8-9. Production chart: 24L dredge for various materials.

Fig. 8-10. Cutterhead dredge with spud barge to increase swing width and dredge efficiency.

SUMMARY

The production of a conventional dredge is limited by suction conditions, pump horsepower, and slurry velocity. The judicious application of the Dredge Laws can lead the way to increased production by identifying the bottleneck and indicating the equipment or change required to break it.

Chapter 9

THE DREDGE CYCLE

The first eight chapters have expounded upon the fundamentals of the hydraulic functions of a dredge. It is now useful to view in simple graphical form the fluctuations through which the dredge passes hydraulically when performing on the job.

A centrifugal pump accelerates the passing liquid to a velocity somewhat greater than the tip speed of the impeller (Chapter 1). This velocity is partially converted to pressure head by the volute. The total generated head is the same for any Newtonian liquid, regardless of its specific gravity. However, if the specific gravity is other than 1.0, head must be multiplied by specific gravity in order to convert to equivalent water head, the unit normally used in dredge calculations. Fig. 9-1 shows the dredge cycle in simplistic form with head expressed in feet of water.

HEAD-CAPACITY CURVE ON WATER

Line E–B is a portion of the pump head-capacity curve on water. This curve is normally provided by the pump manufacturer, and the curve shown is typical for a given RPM. As long as the pump is running at the given RPM, the head and GPM will meet at some point on this curve. It cannot deviate from this curve while pumping water unless RPM is changed.

SYSTEM RESISTANCE ON WATER

Curve ABC is the system resistance on water, representing the flow resistance of the system's line size and length at varying rates of flow. At zero flow, the resistance is zero (point A). As the GPM rises, the resistance rises as a function of the velocity squared

THE DREDGE CYCLE

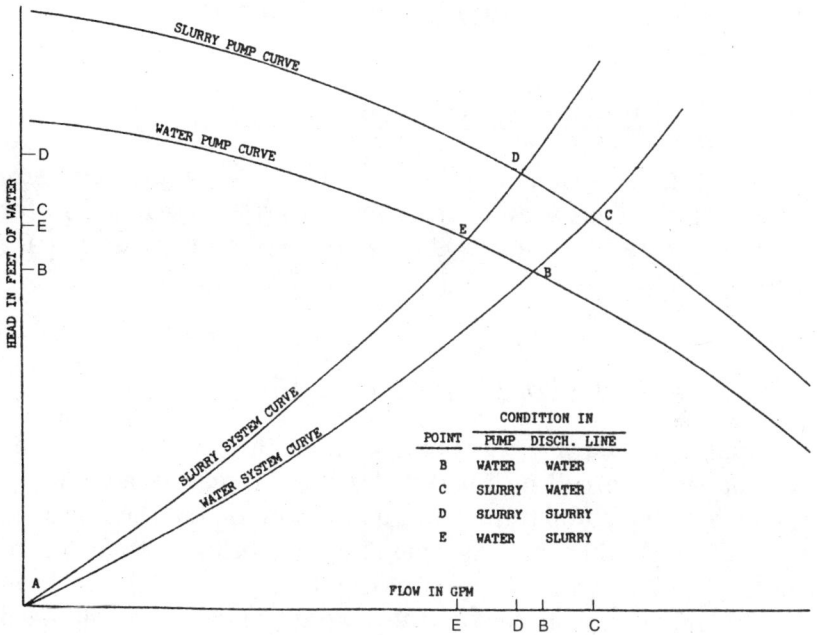

Fig. 9-1. The dredge cycle.

($h \sim V^2$). At point B, the system resistance curve intersects pump head-capacity curve, establishing the point at which the pump will force water through the system. If the line were lengthened or reduced in diameter, curve ABC would become steeper and would intersect the head-capacity curve at a point representing a higher head and a lower GPM.

HEAD-CAPACITY CURVE ON SLURRY

Curve DC is the pump head-capacity curve when operating on slurry with a specific gravity greater than 1.0. Chapter 1 explained that the pressure created by a centrifugal pump is a direct function of the specific gravity of the fluid, but mitigated by the non-Newtonian effect of the slurry. With the head of the pump on slurry expressed in feet of water head (note the ordinate label), the pump head-capacity curve on slurry is above that for water.

SYSTEM RESISTANCE ON SLURRY

Curve AED is the system resistance on slurry. For a given line size and length, the resistance to the flow of slurry is greater than for water; therefore, the slurry system curve is steeper than that for water.

DREDGE CYCLE EXPLAINED

The dredge cycle can be examined within the constraints of the four curves described on Fig. 9-1. Point A represents an inactive system, with the pump turned off. At any point on the water system curve above A, the pump is operating and being brought up to speed. At point B, the pump is at the speed for which the head capacity curve was plotted, and the system functions at the head and capacity indicated on their coordinates by B.

Note that at point B, only water is in the pump and discharge line. Now the operator lowers his cutter into the bottom, starts to swing the dredge, and picks up solids. Abruptly, slurry reaches the pump, which, at the same RPM, produces a greater pressure by shifting to point C on the higher specific gravity slurry pump curve. Since only water is in the discharge line, the GPM and water head immediately increase from B to C.

As the pump continues to force slurry into the discharge line, the resistance of the line increases along curve CD (it cannot leave this curve while at the same RPM on a given slurry) until the line is full of slurry at point D. At this point the system head has increased and the GPM decreased from C.

Point D occurs at the completion of a dredge cut or swing; next, the swing is reversed and the cutter passes over an area already excavated. This results in water at 1.0 specific gravity entering the pump, and, abruptly, the pump performs at point E on curve EB (water). With less head to induce velocity, GPM drops because the system is still providing the high resistance of a discharge line full of slurry.

Fig. 9-2. Dredge *Leonardo DaVinci*, a powerful European 36-inch cutterhead dredge, with rare self-propulsion. Courtesy: IHC.

As water displaces the slurry in the discharge line, the operating point moves along the pump head-capacity curve EB, reflecting a decreased resistance in the line and an increased flow. At point B, the system is at its starting point, where only water is in the system.

Obviously, there are many variations of the dredge cycle, e.g., the percent solids in the slurry can be anywhere between zero and maximum, the discharge line retention time may be longer than the dredge cycle, and the line therefore would almost never be 100 percent slurry or 100 percent water. If, however, we assume the slurry pump curve on Fig. 9-1 represents the performance on maximum percent solids slurry, then the envelope as represented by ABCDE provides the outer constraints of the system. Infinite variations within this envelope are possible.

The reader will possibly detect an element of perversity in the dredge cycle, i.e., the highest velocity occurs at point C when the discharge line is full of water and high velocity is unnecessary; conversely, the lowest velocity occurs at point E when the line is full of slurry, and high velocity is essential to prevent deposition. Without some form of velocity control, the dredge will burn fuel unnecessarily for pumping excessive water, or will jeopardize production by having too low a velocity on slurry. See Chapter 17 for a discussion of velocity control.

SUMMARY

The dredge pump creates head in proportion to the specific gravity of the fluid it impels, while the pipeline frictional resistance, h_F, increases with fluid specific gravity. Since the slurry specific gravity handled by a dredge varies from 1.0 to about 1.5, the dredge cycle passes through broad variations which require understanding by the operator in order to optimize his operations.

Chapter 10

FLOW REGIME AND FRICTION

FRICTION HEAD LOSSES

One of the most difficult problems for the dredgeman is the prediction of friction losses. When one considers the various flow regimes in the slurry pipe; the nature of the solids which vary from flocculates through Bingham plastics to noncohesive gravel; the solids concentrations which cause the slurry specific gravity to vary from 1.0 to 1.6; then one understands why Herbich[3] states:

> It is therefore not surprising that no simple theory and no clear-cut methods are available for estimating slurry head losses for engineering purposes.

Considerable time has been spent developing mathematical models for slurries with a single particle size, but the practical value of such models to the dredgeman is negligible since he never encounters such materials. While many studies have been made of slurry rheology, the infinite diversity of the slurries encountered by the dredgeman responds better to the extrapolation of actual test data and operating experience than to the academic formulas. However, when data is lacking, formulas may represent the best information available, and indeed may be essential.

FLOW REGIMES

Of paramount importance to friction head losses is the flow regime within the pipe. Herbich[3] has defined the four identified slurry flow regimes as homogeneous, heterogeneous, moving bed, and stationary bed.

Fig. 10-1 shows the pipe cross-section for the first three flow regimes above. The homogeneous regime is high velocity, and un-

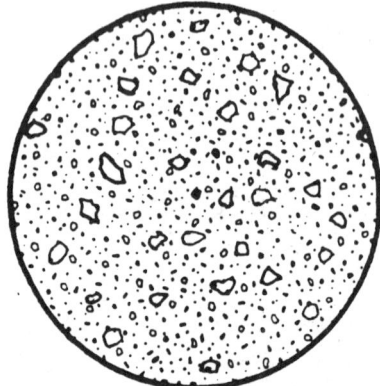

(a) Uniform distribution of sediment. Excessive power requirements.

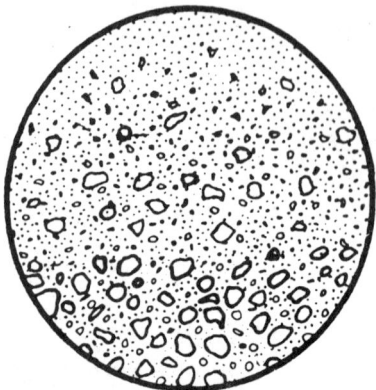

(b) Nonuniform distribution of sediment. Optimum power.

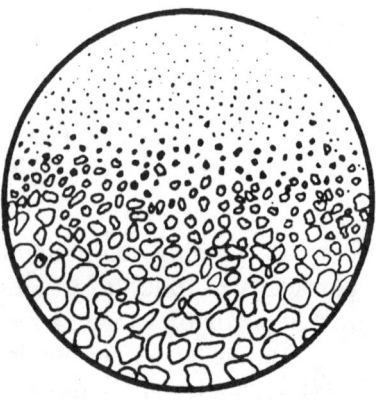

(c) Segregated distribution of sediment. Bed moving at lower velocity than material above. Excessive power requirements.

Fig. 10-1. Sediment distribution in a pipeline. (a) homogeneous flow; (b) heterogeneous flow; (c) flow with a moving bed. Reprinted, by permission, from Herbich, *Coastal and Deep Ocean Dredging*, 297.

economic in terms of power and wear. The heterogeneous regime is the most economic for solids transport. The moving bed regime reduces production and is uneconomic, but is not uncommon on dredges. The stationary bed is uncommon because of its low productivity, and should be avoided. The heterogeneous regime should be used when possible. This desirable regime occurs in the vicinity of the 1.5 specific gravity optimums shown on Fig. 4-1 as a function of velocity and the nature and percent of the solids.

HEAD LOSS CALCULATION FROM OPERATING DATA

The forward-thinking operator should always be aware of the material he is pumping and log his conditions so as to be able to calculate his frictional losses per 100 feet of equivalent line length. When frictional aberrations occur, they should be analyzed as to cause, e.g., shell, gravel, clay, etc. It is a relatively simple matter to calculate the loss per 100 feet, but in order for it to have significance, the velocity, solids concentration, and solids sieve analysis must be known. Using the discharge pressure gauge reading, the following equation applies:

$$2.31P + h_G = h_F + (h_{TE} \times SG_s) \quad [Equation\ 10\text{-}1]$$
$$\text{Or } h_F = 2.31\ P + h_G - (h_{TE} \times SG_s)$$

Where P = gauge pressure in psi
h_G = feet of gauge above water level
h_F = total friction head in feet of water
h_{TE} = terminal elevation of pipe discharge in feet above water level

Assume that the SG of the slurry is 1.25; the equivalent line length is 4,000; the gauge is 20 feet above water level and reads 80 psig*; and the terminal elevation is 10 feet; then calculate as follows:

$$2.31 \times 80 + 20 = h_F + 10 \times 1.25$$
$$204.8 = h_F + 12.5$$
$$h_F = 192.3 \text{ total friction head in feet of water}$$
$$h_F/100 \text{ feet} = \frac{192.3}{4{,}000} \times 100 = 4.81 \text{ ft}/100 \text{ ft}$$

No theoretical formula can compete with the accuracy of this simple calculation. However, the dredgeman is faced with the necessity of predicting total friction head, h_F *prior* to actual operation since he must submit a bid in advance which includes the necessary pumps and boosters. It is therefore recommended that the operator, in preparing his bids, refer back to his prior operating experience for similar material and conditions for the best prediction of his future performance.

---

*Note that if the gauge is read in the lever room, the elevation correction, h_G, may be roughly equivalent to h_{TE}; so many operators simply equate h_F to gauge reading $\times$ 2.31.

In the event there is no appropriate prior experience, there is still good information available. For sands and gravel, there is much data and experience on which the production curves and limiting velocity curves (Chapter 4) are based. It should be noted that these materials are classified by their median grain size, d_{50}, and are ungraded, noncohesive, and have little or no clay content.

SOIL TYPES

As far as the rheology of dredge hydraulic transport systems is concerned, there are three broad classes of soils: cohesive, noncohesive, and mitigated.

Cohesive materials are soils with a fines content (usually clay) of such affinity that the soil does not separate and disperse readily in the slurry. *Noncohesive materials* are sands and gravels (with little or no clay or other cohesives) which disperse readily into discrete granules. *Mitigated materials* are those which are made up largely of noncohesive materials, but whose rheological characteristics are mitigated by a percentage of clays and/or silts. These mitigating clays have the effect of increasing the specific gravity of the conveying medium; of providing some aspects of what Stepanoff[13] called "plug flow" regime; and, in effect, acting as a lubricant for the system. Many operators have been pleasantly surprised by the reduction in friction loss for a material with some clay or other mitigating fines.

Huston[2] suggests in calculating slurry friction losses that water losses be computed, and then multiplied by the specific gravity of the slurry. This is a reasonable procedure on cohesive soils such as the silt and mud found in many maintenance projects, and on mitigated soils where the system is "lubricated." However, test data and actual dredge operations have disclosed instances where the water multipliers exceed 2.0, so that the specific gravity multiple is not universally applicable.

C FACTORS FOR HAZEN-WILLIAMS EQUATION

The use of specific gravity as a water friction multiplier fails to consider the different friction losses of fine, medium, and coarse sand and gravel, and is therefore limited in its application. For

FLOW REGIME AND FRICTION

accuracy in calculating friction losses of various materials at varying concentrations, it is necessary to have a proven equation and a guide for selecting the friction factors to be utilized with the equation. The Fanning equation, Equation 6-2, is used by some, but more favored is the well-tested, empirical Hazen-Williams equation as follows:

$$f = 0.2083 \left(\frac{100}{C}\right)^{1.85} \times \frac{q^{1.85}}{d^{4.8655}} \quad [Equation\ 10\text{-}2]$$

$$\text{or } f = 1.09 \left(\frac{100}{C}\right)^{1.85} \times \frac{V^{1.85}}{d^{1.1655}} \quad [Equation\ 10\text{-}2A]$$

Where f = slurry friction loss per 100 feet of pipe expressed in *feet of water*
d = inside diameter of pipe in inches
q = GPM
C = friction factor taken from Fig. 10-2

Velocity (or GPM) and pipe inside diameter in inches must be known, and "C" determinable in order to calculate f.

Fig. 10-2 shows estimated "C" values as a function of d_{50} (median grain size) and slurry specific gravity. This chart has been developed from practical data acquired over the years from many

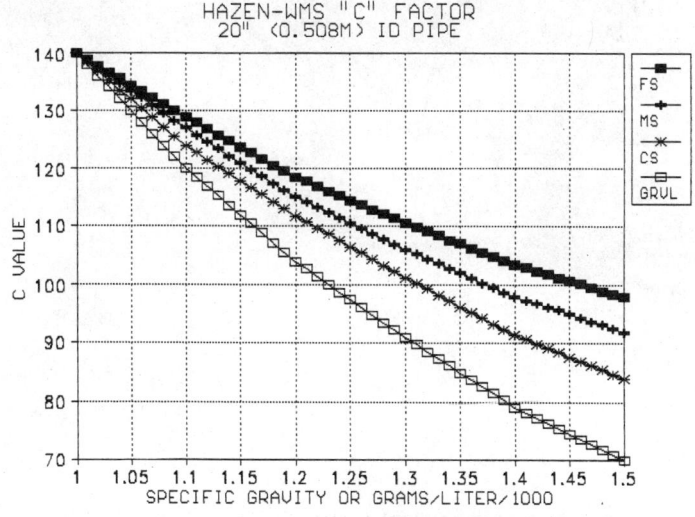

Fig. 10-2. Hazen-Williams "C" friction factor.

sources. It is presented, not as a precision instrument (since laboratory data is unavailable), but as a highly useful tool in the prediction of friction losses in the absence of better data. There may be instances where d_{50} on the chart will not truly reflect the rheology of the slurry because of abnormal grain distribution; or where a lubricating clay mitigates the rheology to *increase* the "C" value by perhaps 10 points, thus *reducing* losses. Even with possible inaccuracies, this chart, Fig. 10-2, along with the Hazen-Williams equation (soluble quickly with the common hand-held scientific calculators) constitutes a good method of prediction of friction losses, perhaps as good as any method, other than that of actual operating data on similar material.

It should be emphasized that use of "C" values from Fig. 10-2 results in friction losses expressed in *feet of water*, not slurry. For this and other reasons, it is acknowledged that some operators use "C" values that differ from those above. The operator should cling to his own data if it works for him, but remain flexible and rec-

Fig. 10-3. 24-inch Australian mineral mining dredge with bucketwheel excavator. Courtesy: Ellicott Machine Corp.

ognize that h_F varies with type and percent of solids (from zero to 50 percent) and size and shape of solids (from microns to cobbles).

It is incumbent upon each operator to adopt a method of calculating friction that allows him to predict losses with accuracy, and thus to bid with confidence. The Hazen and Williams friction formula used with Fig. 10-2 has been used with good success. Computerized calculation is recommended.

SUMMARY

Slurry friction losses vary with the type and concentration of solids being pumped, as well as the flow regime within the pipe. The Hazen-Williams friction factor, used in conjunction with Fig. 10-2 and a personal computer, is perhaps the most satisfactory way to predict friction; however, good historical data has been used successfully by operators who were alert enough to ensure that similar conditions were being compared.

Chapter 11

CAVITATION

There are no good results from dredge pump cavitation. It disrupts the dredging function and can cause equipment damage. It is essential that the operator learn to minimize cavitation.

Much has been learned and written about the theory of cavitation; however, the dredge operator's interest lies not so much in the arcane theory, as in being able to predict and avoid the circumstances that cause it. This chapter is devoted to that end.

DEFINITION

The cavitation phenomenon occurs when the static pressure at the impeller eye of the pump falls below the vapor pressure of the liquid being pumped. This results in liquid (normally water) vaporization, which forms low pressure "cavities" in the slurry. These cavities later implode, often with audible effect, causing physical stress and potential damage to the pump metal in the area of the implosion.

Perhaps the simplest concept of cavitation is to visualize the slurry entering the rotating impeller through the eye. Each impeller vane terminates at the periphery of the eye at the front shroud. As long as there is adequate suction head to keep the slurry in close contact with the trailing surfaces of the moving vanes, all is well; however, as RPM increases and vane velocity at the eye periphery (commonly called eye speed) exceeds the ability of the slurry to keep pace, the vane "runs away" from the slurry, forming cavities of water vapor, i.e., cavitation. Unless the dredge is equipped with a method to augment the suction head, (normally supplied by barometric pressure only) the dredge will inevitably encounter cavitation.

CAVITATION

When cavitation occurs, the pump tries to pump the water vapor, a fluid lighter than air. Since the discharge pressure created by the pump is a function of the specific gravity of the fluid being pumped (Chapter 1), the pressure created with 100 percent water vapor is so low as to be unmeasurable by normal dredge instruments. The practical effect of full cavitation is a cessation of liquid pumping. This not only stops dredge production (i.e., the transport of solids), but allows settlement of the suspended solids in the slurry, resulting in potential choking of the pipe.

CAVITATION CHART

The graph of Fig. 11-1 is a reasonable approximation of the capabilities of a well-designed dredge pump. It shows the relationship between the pump's eye speed and the maximum vacuum the pump can create. Eye speed is the peripheral velocity of the opening through the front shroud of the pump impeller, expressed in feet per second. Vacuum is "negative pressure," a misnomer reflecting the extent to which the absolute suction pressure is reduced below barometric pressure. It is read by a gauge immediately ahead of the pump, and is traditionally expressed in inches of mercury (feet of water would be preferable for ease in calcu-

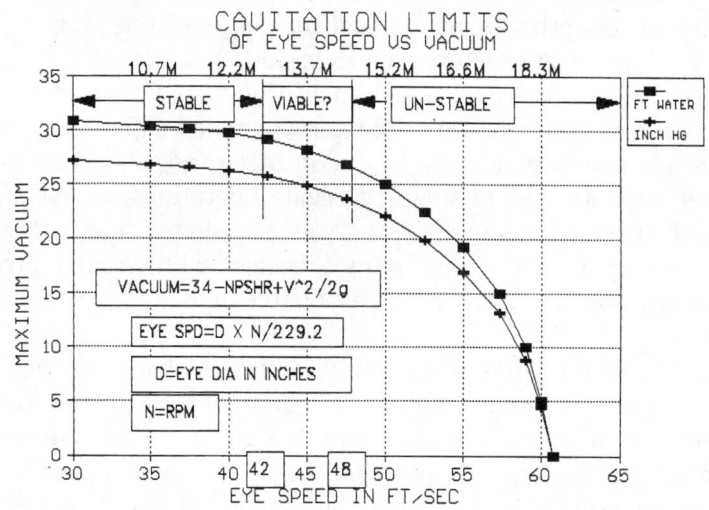

Fig. 11-1. Cavitation limits of eye speed vs. vacuum.

lations). From Fig. 11-1, it is obvious that as eye speed increases, vacuum capability decreases. Since required vacuum increases as a function of slurry velocity and specific gravity (SG), it becomes apparent that the productivity and economics of the dredge depend upon the ability of the pump to create a high vacuum.

EYE SPEED

Examination of the graph discloses that at an eye speed of 30 ft/sec, the pump can create a vacuum of 27-inch mercury (Hg). At 42 ft/sec, the vacuum has dropped a small amount, so the question arises, "should the pump always operate at a maximum eye speed of 30 ft/sec?" The answer is "no." As the eye speed is reduced, so is the peripheral tip speed of the impeller, which determines the head created by the pump. Since head is a function of the square of the tip speed, this means that at an eye speed of 30 ft/sec, the impeller tip generates a head only 49 percent of that at 42 ft/sec, and thus will pump only about 49 percent as far. Also, since flow varies directly as the pump speed, the GPM will be only 71 percent as high. Further, pump HP varies as the cube of RPM, so the HP utilized will only be 36 percent of that at 42 ft/sec.

Obviously, the dredge designer must optimize the pump speed and HP for the project conditions to be encountered. The complexities of this process are beyond the scope of this chapter, but a good compromise for eye speed is 40–42 ft/sec, at full prime mover speed and HP. Perhaps an inch of vacuum is lost versus 30 ft/sec, but this is a small price for the increased capability in flow rate, head, and productivity of a contract dredge. On short-line dredges, such as used in some aggregate operations, an eye speed of 30–40 ft/sec may be appropriate when using a standard, commercial pump; however, on contract dredges with variable project conditions, the 42 ft/sec is a reasonable and economic design criterion.

Further examination of the graph discloses that at about 61 ft/sec eye speed, the pump has no vacuum capability. Long before this point is reached, however, the dredge pump has passed its maximum practical operating speed, because it must generate substantial vacuum in order to pick up solids, to generate velocity head, and to overcome the losses in the suction line. The previ-

CAVITATION

Fig. 11-2. Great Lakes 30-inch dredge, *Alaska*, with ladder pump and anchor booms. Courtesy: Mobile Pulley & Machine Works.

ously stated 42 ft/sec is a good eye speed for a dredge pump; however, if the vacuum requirements never exceed 15-inch Hg (e.g., a booster pump) the eye speed can be increased to about 55 ft/sec to take advantage of the longer line length capability (assuming adequate HP availability). Also, a dredge pump with an eye speed of 45–50 ft/sec can function on long lines by lowering the vacuum to pick up less solids; however, the HP of the prime mover must be increased to handle the higher speed, which means that under "normal" project conditions, the available hp is not utilized. Similarly, for a dredge often pumping against long lines, an eye speed up to 45 ft/sec is not unreasonable; however, if a booster pump is used, the operation would be more effective if the dredge pump were designed for a maximum of 42 ft/sec.

NET POSITIVE SUCTION HEAD REQUIREMENT

Operators have noted that their dredge pump does not always cavitate at the same vacuum. The graph, with its formula for vacuum, demonstrates this. The formula states that vacuum expressed in feet of water equals barometric pressure (assumed as 34 feet) minus the net positive suction head required (NPSHR) plus the velocity head (velocity squared divided by 64.4). NPSHR is a concept not essential to the average operator, so we can dismiss it by recognizing it as a calculated value of suction head required for the pump to operate under the given conditions.

NPSHR is normally determined by the pump manufacturer from tests, and should be provided with the pump curves. NPSHR increases with eye size, and therefore it is recommended that eye diameter never exceed the inside diameter of the suction pipe. Some operators have increased eye diameter in an effort to pass larger particles through the pump, but this often comes at a high operating price and is seldom, if ever, the most effective way to achieve the objective.

IMPELLER GEOMETRY AND SPEED

Most dredge pumps will generate a maximum head of about 240–260 feet with a vane tip speed of 113 ft/sec. With an eye speed of 42 ft/sec, this provides a ratio of 113/42 = 2.69. This ratio of impeller diameter to eye diameter normally provides satisfactory performance if the pump speed and HP are properly matched. An increase in tip speed allows longer lines (assuming the eye is diminished to avoid cavitation), but requires higher HP, and results in higher wear. A decrease in tip speed has the reverse effect.

SUMMARY

Cavitation can devastate the performance of a hydraulic dredge. It behooves the operator to obtain well-designed pumps with matched drives, and to operate within their capabilities. He should also recognize the following principles:
1. The ability of a dredge pump to create vacuum is reduced as eye speed increases.
2. Normal impeller eye speed should not exceed 42 ft/min (12.8 m/min).
3. The eye diameter should not exceed the inside diameter (ID) of the suction line; it is acceptable practice to make it the size of the discharge line ID.

Part II
DREDGING IN PRACTICE

Dredge *R. N. Weeks* filling its hopper. Courtesy: Weeks Marine Inc.

Chapter 12

SELECTING THE DREDGE TYPE

Hydraulic dredges are characterized by the use of a centrifugal pump to induce a high velocity water stream in a pipeline in which solids are entrained and transported to their discharge area. They are further categorized by their method of excavation, i.e., by the nature of the intake element in contact with the dredged material. The major types are plain suction, trailing suction, and cutterhead.

PLAIN SUCTION

This is the simplest form of hydraulic dredge and uses no excavator. It is sometimes equipped with water jets to agitate the dredged material to facilitate solids entrainment by the water entering the suction mouth. This dredge is limited in application to relatively soft and free-flowing materials, and is not easily adaptable to digging channels since it cannot be swung across the channel continuously without danger of structural failure. It is used primarily for "winning" or acquiring material from the waterway by creating a large inverted cone in the bottom. Its efficiency can be high and its cost per cubic yard low when properly applied, but its application is severely limited by its inability to excavate, to dig a channel, or to mine a horizontal stratum.

The dustpan dredge, Fig. 12-2, is a form of plain suction dredge which derives its name from its special suction head (Fig. 12-3) which may be 30 feet in width or greater. It is equipped with multiple jets to agitate the bottom material. The dredge is pulled forward (normally against the current) by crossed wires attached to upstream anchors. It sweeps a broad straight channel, and is perhaps the most effective tool available for the quick removal of shoals. It pulls its discharge line along behind it, utilizing an ingen-

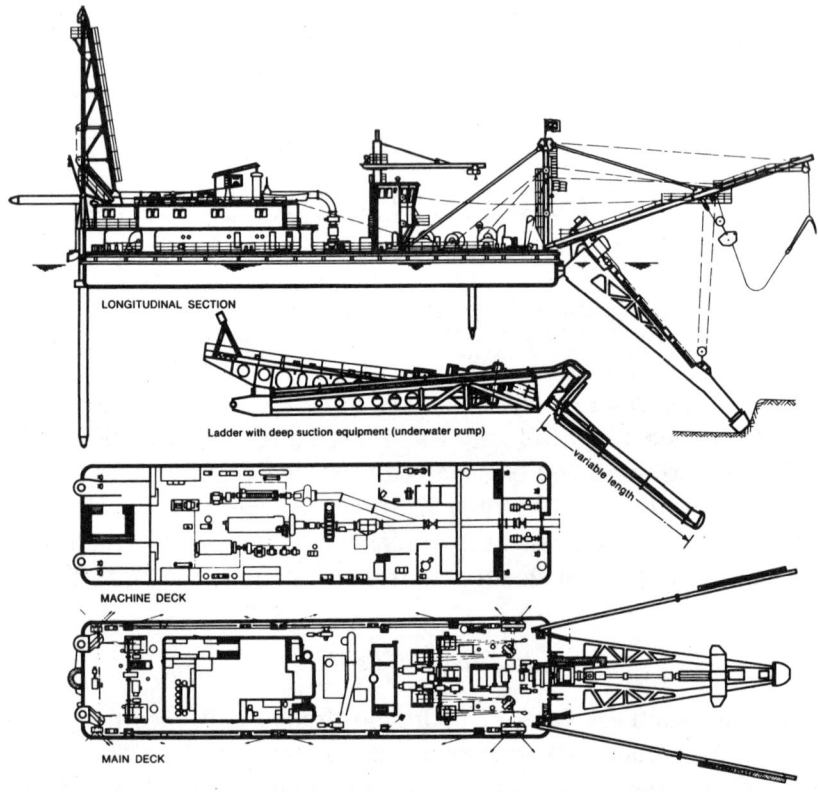

Fig. 12-1. Dredge diagram *Spuler VI* with plain suction convertible to cutterhead. Courtesy: Philipp Holzmann.

ious discharge barge incorporating a baffle against which the pipeline discharges its slurry. The reacting forces position the barge in the shallows or wherever the operator may direct for the deposition of the material. The dustpan dredge has seen limited application in the industry and deserves to be better known for its high capacity and low cost.

TRAILING SUCTION

The trailing suction dredge, Fig. 12-4, is a self-propelled, oceangoing vessel generally compartmented into several hoppers. The most common configuration has two dragarms, one on each side

Fig. 12-2. Dredge *Lenel Bean*, 18-inch dustpan. Courtesy: C. F. Bean Corporation.

of the ship, mounted outboard and connected to the hull near the center of buoyancy to minimize the effect of the sea state.

Other configurations may involve only one dragarm mounted on one side of the vessel, or at the stern on the ship's centerline. Each dragarm has its own draghead for contact with the bottom and with minor exceptions, serves its own separate pump.

Some trailing suction dredges have no hoppers and discharge their loads overboard through extended, cantilevered discharge lines. These dredges are called "sidecasters." The more common "hopper dredge" discharges into its own distribution system which is frequently so versatile as to allow either or both pumps to direct the effluent to any of the several hopper compartments.

While dredging, the vessel is underway at about 2 or 3 knots, with the draghead trailing from the trunnion-mounted dragarm so as to absorb the motion of the hull in the sea state without ill effect. Effluent is pumped into the hoppers where the solids tend to settle to the bottom. After the hoppers are full, overflow to the sea begins. This overflow is water with some solids content as a function of the settling time available. When the economical solids

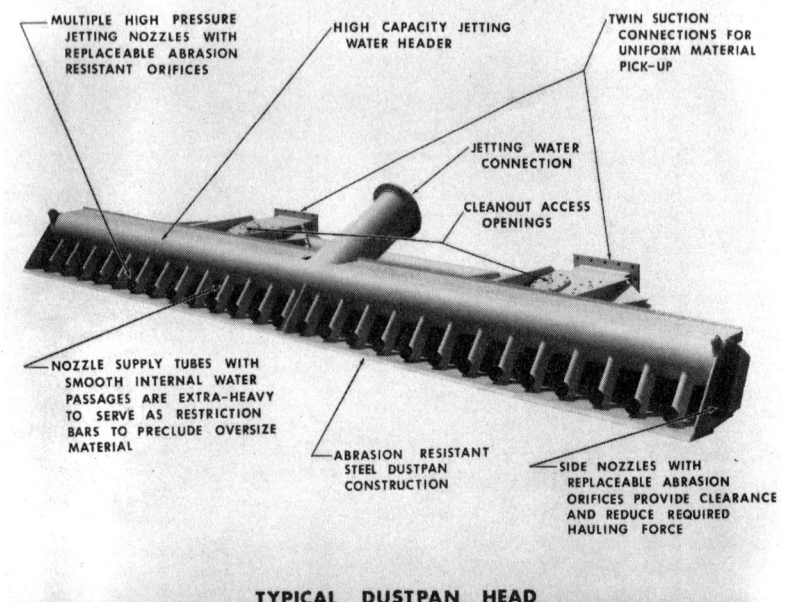

Fig. 12-3. Typical dustpan head. Courtesy: Ellicott Machine Corp.

load has been taken on, the dragheads are elevated, and the ship proceeds to the dumping ground, frequently in deep water, where the bottom hopper doors open and the load is discharged. The dredge then returns to the dredging grounds for another load.

Hopper dredges were developed for maintenance work, the first being the *General Moultrie* in 1855 for work on the Charleston, South Carolina, bar. They were intended for soft or free-flowing materials. However, with the appropriate draghead, Fig. 12-5, they have proved capable of dredging surprisingly difficult virgin material.

Hopper dredges are advantageous in busy channels or harbors where traffic and operating conditions preclude the use of stationary (swinging) cutterheads with their attendant pipelines. They are also capable of operating in a sea state of several feet which would inactivate the normal cutterhead dredge. They can mobilize quickly since they are able to proceed under their own power, and require less in the way of support craft than the cutterhead. They achieve rapid improvement in a channel by traveling the full length of the

SELECTING THE DREDGE TYPE 105

Fig. 12-4. Split hull hopper dredge. Courtesy: Twin City Shipyard, Inc.

channel while not blocking it, whereas a mechanical or cutterhead dredge proceeds laboriously in perhaps 3- to 6-ft deep cuts across a wide channel, effectively blocking a portion or all of the channel. Similarly, the hopper dredge can excavate deep cuts the full length of a shoal so as to concentrate the current flow and induce scouring. They are also advantageous where the dumping grounds are unavailable for a cutterhead, or are so distant as to be uneconomical to pump. These advantages are such as to probably always assure the existence of a number of hopper dredges.

The hopper dredge has disadvantages; it is built to satisfy "ocean" classification and is therefore quite costly. It requires manning in accordance with oceangoing marine practices, also costly. Further, when transporting its load to the dumping ground, it is certainly one of the most expensive dump-scows ever devised. It cannot dredge irregular patterns, operate near piers or other obstructions, or operate in shallow water; neither can it dredge some hard materials successfully, which can be dug by cutterheads. It

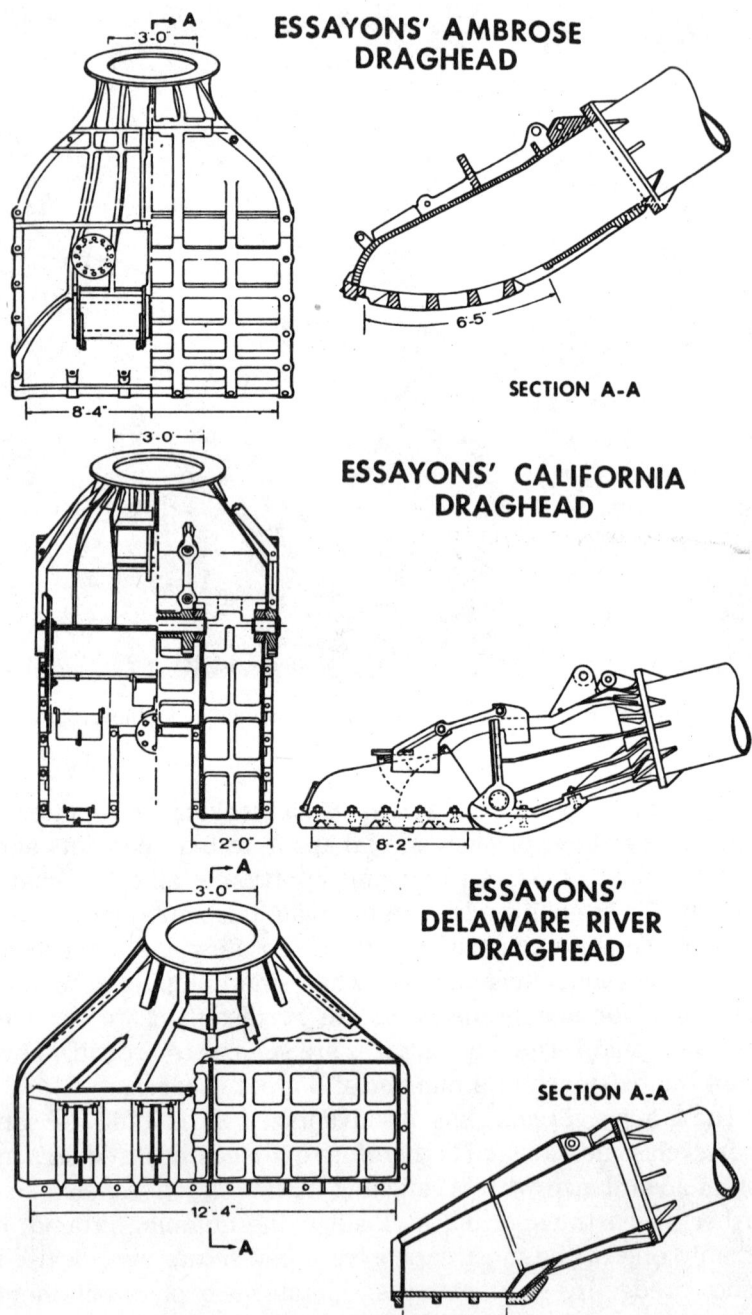

Fig. 12-5. Dragheads.

also requires double handling of material whenever dry land deposition of the dredged material is required. These disadvantages assure that the hopper dredge will always be supplemented by other forms of dredges, particularly the cutterhead.

CUTTERHEAD

The most common and most versatile hydraulic dredge is the cutterhead, Fig. 12-6, which is equipped with a rotating cutter (excavator) surrounding the intake of the suction line. The cutter excavates and translates the bottom materials into the influence of the high velocity water at the suction intake, where the solids are entrained, passed through the dredge pump to the floating discharge line, and on to the deposition area through the shore pipe.

The conventional cutterhead dredge is held in position by two spuds at the stern of the dredge, only one of which can be down (i.e., in the bottom) while swinging. There are two swing anchors some distance from either side of the dredge which are connected

Fig. 12-6. A 24-inch cutterhead dredge. Courtesy: Ellicott Machine Corporation.

by wire rope (through swing sheaves on the ladder mounted near the cutter) to the swing winches. The dredge swings to port and starboard alternately, while passing the cutter through the bottom material until the proper depth is achieved. Most cutterheads advance by "walking" themselves forward on their spuds. This is accomplished by swinging the dredge to the port on the port spud an appropriate distance. Then the starboard spud is dropped and the port spud raised. When the dredge is swung an equal distance to the starboard, the port spud is dropped and the starboard spud is raised. The dredge has now advanced a cutterhead length and is on course, if the proper advance angles are used.

The cutterhead dredge has most of the advantages of the other type dredges, and can perform on a continuous basis with resultant improvements in efficiency and costs. It can operate in an identical fashion to the plain suction dredge, but can also dig a specified channel including slopes as required. It can dig the same channels and with more precise control than the trailing suction dredge, while discharging to an upland disposal area. If the disposal area is distant, several miles or more, the cutterhead can discharge into barges for economical transport, not interrupting the dredging cycle as with the hopper dredge. Its great versatility, continuous operation, and moderate costs have earned it its appellation as the workhorse of the industry.

The cutterhead dredge does have disadvantages, however, and they are its inability (1) to handle large particles such as boulders, slabs, stumps, etc., and (2) to work under heavy sea conditions. Particle size limitations will be discussed in Chapter 13 under pump design, but it should be noted that proper clearing and/or blasting of the area to be dredged, along with proper design of the pump impeller, goes a long way toward removing this disadvantage. As for coping with a heavy sea state, there are several schemes to adapt the cutterhead dredge to heavy seas, including patent #3,777,376 issued to this author and associates.

COMPENSATED CUTTERHEAD DREDGE

Fig. 12-7 shows the patented, compensated cutterhead dredge, adjustable to cope with varying sea and soil conditions. This dredge was intended to operate successfully in seas of at least two meters.

SELECTING THE DREDGE TYPE

Fig. 12-7. Self-compensated cutterhead dredge. Courtesy: Ellicott Machine Corporation.

It has an articulated, compensated ladder assuring high dredge efficiency, and preventing the dredge's destruction when operating in open water. The principle of a gas-charged, hydraulic compensating device has been proven on floating drill rigs. Its unique application to the cutterhead dredge can be seen in this artist's conception. Note that the hull is more or less conventional, but the ladder and attendant forward frames are novel. The ladder is articulated with the conventional trunnions at the hull, but with two additional trunnions just ahead of the cutter module. The need for two additional trunnions is clear from model tests which indicated that only one additional trunnion would allow destruction of the mechanism when the cutter and hull trunnion were in a straight line with the additional trunnion.

To understand the compensating principle, assume the weight of the cutter module is 100,000 pounds and that all of this weight is carried by the hydraulic cylinders and the soil reaction. Now assume that the soil requires 50,000 pounds cutting force leaving 50,000 pounds for the cylinders, easily adjustable by the operator. As the dredge rises on a swell, the cylinders will lift only 50,000 pounds; this causes the cylinder rods to extend, in order for the soil resistance to supply the other 50,000 pounds. As the dredge falls the rods will retract, reversing the procedure. The cutter remains in constant contact with the bottom, assuring high dredge efficiency, while the entire ladder is protected against the heave, roll, and surge of the hull.

SUMMARY

The type of hydraulic dredge selected for a project is a function of the project condition and requirements. If the operation is primarily a sand winning operation to provide fill, and contour of the bottom is unimportant, the plain suction dredge may apply. If the estuary to be dredged has a specified channel and heavy marine traffic, the hopper dredge may be applicable. Large, loose shoals can be efficiently removed by dustpan dredges. Under any of the above conditions, the versatile cutterhead dredge may prove to be the advantageous selection.

The cutter.

Chapter 13

THE CUTTER

TYPES AND FUNCTIONS

The cutter of the cutterhead dredge is the excavator which surrounds the suction intake. It has two primary functions: (1) to loosen and disintegrate the bottom material into particle sizes compatible with the pumping system; and (2) to place the disintegrated material in the high velocity stream at the suction intake in the necessary quantity.

Most cutterhead dredges use the traditional basket cutter in one or more of its variations; however, there are three other types of cutters that have achieved some popularity. They are the bucket wheel, the endless chain, and the high-speed disk.

THE BASKET CUTTER

This traditional cutter is a multibladed excavator which rotates around a longitudinally mounted shaft. Figs. 13-1 and 13-2 show it in its simplest, single-piece casting form, as well as in complicated fabrications involving replaceable edges and hardened, pinned-on teeth. The basket cutter can vary in shape, hand, number of blades, type of cutting edge, method of attachment, rake angle, etc., and since there is little hard engineering data available to the industry regarding the effect of the various designs, there is a wide divergence of opinion in the industry as to their value.

The spider cutter is a popular variation of the basket (see Fig. 13-3). The true basket curves the crown end of each blade back into the drive hub for support. The spider terminates the blades in space, achieving support by extending short arms from the hub to an intermediate section of the blade.

 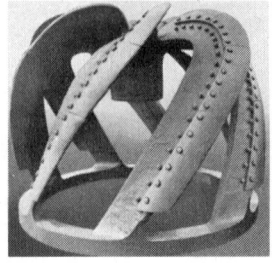 

Fig. 13-1. Plain basket cutters. A: Single casting, 6 blades; B: Bolted edges, 7 blades; C: Fabricated, welded edges, 8 blades. Courtesy: Ellicott Machine Corporation.

Fig. 13-2. Toothed basket cutters. A: Welded serrated edges, 6 blades. Courtesy: Mobile Pulley, Inc. B: Crude welded teeth, 6 blades. Courtesy: Ellicott Machine Corporation. C: Crude pinned teeth, 6 blades. Courtesy: Ellicott Machine Corporation. D: Pinned teeth, welded bi-leg adaptor. Courtesy: Mobile Pulley, Inc. E: Pinned teeth, round base adaptor. Courtesy: Mobile Pulley, Inc. F: Pinned teeth, renewable edge. Courtesy: Mobile Pulley, Inc. G: Pinned teeth, renewable edge. Courtesy: Ellicott Machine Corporation.

THE CUTTER

Fig. 13-3. Spider cutters. A: With renewable blades. B: With renewable blades with trash bars. Courtesy: Ellicott Machine Corporation.

On granular, free-flowing material, almost any moderately sturdy basket design is successful. As long as the cutter runs interference for the suction inlet, i.e., prevents excessive stresses on the suction by displacing the material and bringing it inside the cutter within the influence of the high velocity stream of water, it will perform. It should be emphasized that "within the influence" means the suction inlet should be essentially in contact with the material or even buried. Note the illustration in Fig. 13-4 which shows a pipe representing a suction inlet. At the vertical plane of the inlet, the pipe area is πR^2. If the pipe is 24 inches, the limiting velocity curve, Fig. 4-1, shows that 16 ft/sec velocity is required for

THE CUTTER

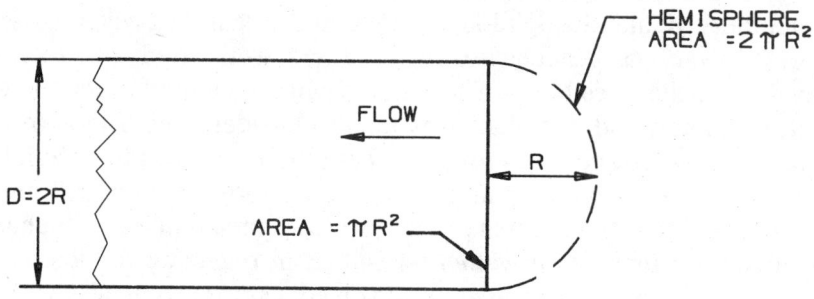

Fig. 13-4. Suction inlet—distance effect.

optimum transport of coarse sand through the suction area. However, if the center of the inlet is one radius away from the material, the area of the hemisphere is $2\pi R^2$, and the velocity only 8 ft/sec at the nearest material, which results in a negligible pickup of solids.

Most problems with cutters are encountered with materials which must be disintegrated or sized so as to be compatible with the transport system. The material may be dense or tenacious, necessitating high force and/or edge velocity to perform the excavation. If the cutter proves adequate for this first function, excavation, it may yet fail in the second, i.e., bringing material into the proximity of the suction inlet. Cutters with the speed to mill away hard material may scatter the resulting particles over the bottom and achieve very low production. The author observed an operation where a specially developed cutter disintegrated limestone by massive application of power and milling action but, even when overdigging by 12 feet, was unable to ingest a profitable payload since it effectively dispelled most material it contacted. After the bottom was pulverized to a 12-foot depth, a reduction in RPM and swing speed of the massive rock cutter might well have increased production. Or, a change to a more open, acquisitive cutter could have been effective. Again, the suction inlet must be in close proximity to, or preferably buried in, the material to carry a good load, particularly with heavy, coarse materials. It is the author's considered opinion that basket cutters fail as frequently in the second function, i.e., putting material in proximity to the suction inlet, as they do in the first function, excavation.

A few years ago, a well-known processor of heavy minerals was failing to achieve desired dredge capacity when encountering hardpan. Capacity would drop by 25 to 35 percent when hardpan was encountered, and the opinion was unanimous that the basket cutter horsepower was inadequate. New cutters were tried, but to no avail, and the possibility of a major change including a complete cutter, motor, and line shaft was under consideration. The author was called in for consultation and noted the velocity in the suction pipe was borderline on the granular sand normally pumped, and recognized that the hardpan, in larger fragments, required a higher velocity. An increase in velocity resulted in regaining the lost capacity, and the cutter problem was solved. Cutter problems should always be considered in context with the hydraulic transport con-

THE CUTTER   117

ditions that prevail since the successful disintegration of the bottom by itself, puts no material on the bank.

Fig. 13-5 shows an artist's idealistic version of a basket cutter in operation. The solid in this drawing is showing amazing cooperation by rushing to the suction inlet for no apparent reason. While the cutter blades are designed to afford some transportation of the material toward the inlet, this is achieved only when the cutter is full. And, a full cutter is achieved only when the cutter is essentially buried in material. A significant shortcoming of the basket cutter is that the trailing half of the cutter is wide open, allowing the excavated material to fall out. Therefore, the basket cutter is capable of picking up perhaps half of the material that it excavates and is not an efficient device. See Fig. 13-6.

Cutter Shape
Fig. 13-7 shows the face angle and cone angle of the basket cutter.[12] When the face angle is zero and the cone angle zero, the cutter is

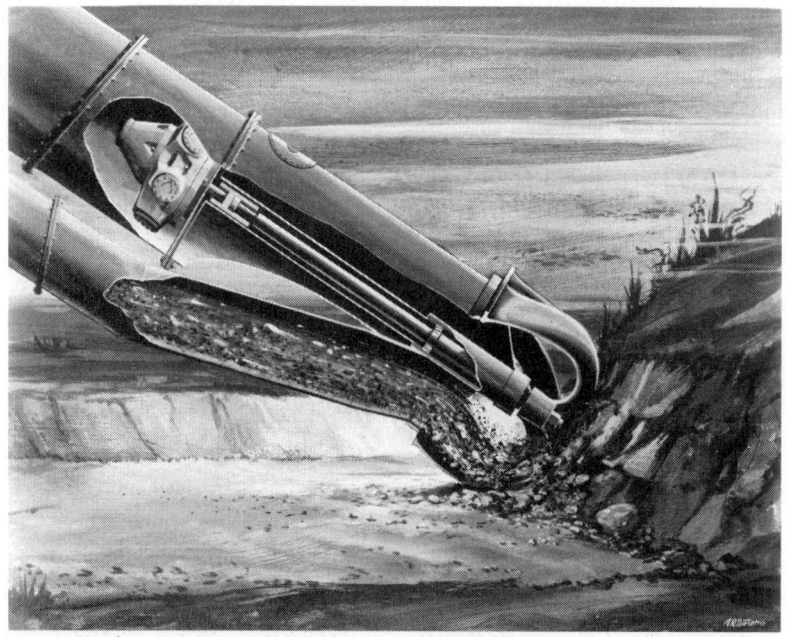

Fig. 13-5. Basket cutter, idealized operation. Courtesy: Ellicott Machine Corporation.

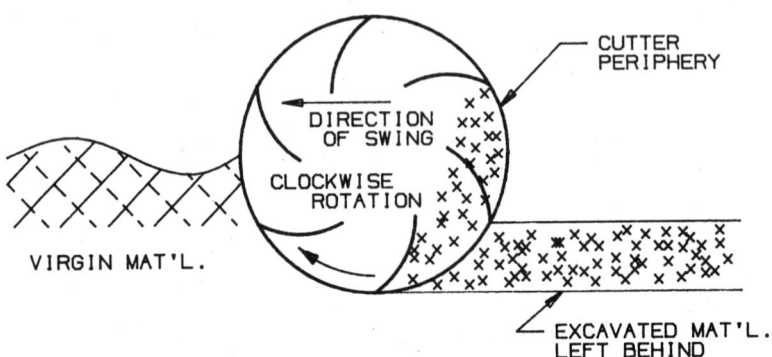

Fig. 13-6. Basket cutter showing retention problem.

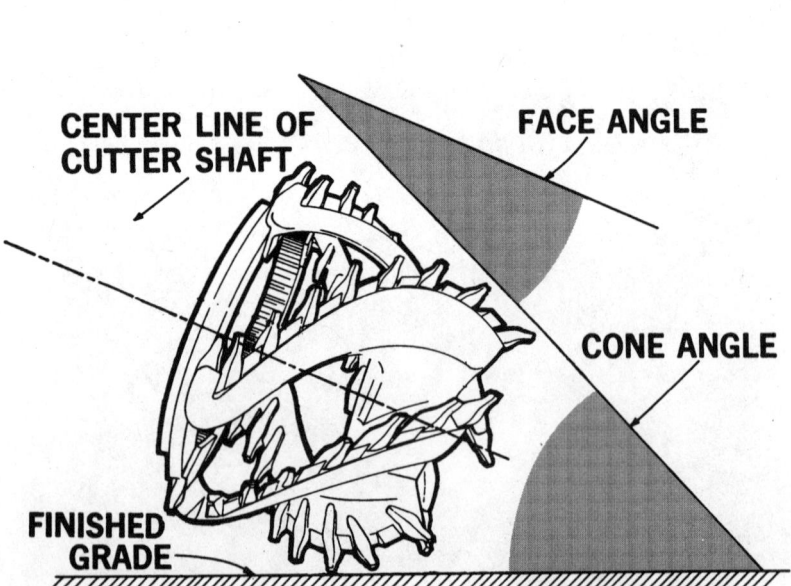

Fig. 13-7. Cutter face and cone angles. Courtesy: World Dredging Conference.

said to be "square" or apple shaped. Optimally, the face angle should vary as a function of the angle of attack of the cutter to the bottom, or more simply stated, as a function of the cutter shaft angle to the bottom. When the cutter shaft angle is small, a large face angle, creating a highly conical cutter, can bring the back ring of the cutter into contact with the bottom. This is not necessarily bad, since the suction inlet is at the back ring, but the stationary ladder structure is immediately adjacent to the back ring, and this condition can lead to "dragging the ladder" or pulling stationary surfaces through the bottom. This causes high line pull, broken swing wires, and low production. A square cutter is more appropriate for shallow digging (low shaft angle), whereas a high face angle (perhaps 20° to 30°) is more effective for deep digging (high shaft angle). Ideally, the cutter should cut a finished grade approximately parallel to the water surface.

Displacement Angle

Fig. 13-8 shows the angle of displacement of a single blade. A greater angle of displacement makes for a smoother operation and

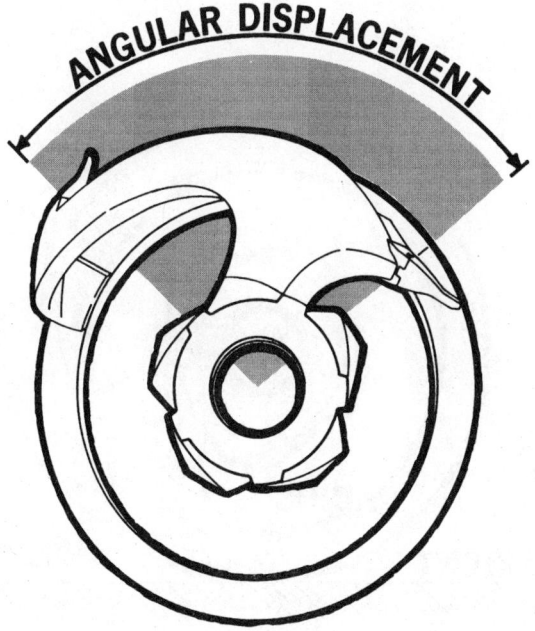

Fig. 13-8. Cutter blade displacement angle. Courtesy: World Dredging Conference.

greater penetrating force. A large displacement angle is particularly important when the traditional six blades are reduced to five or four.

Some cutters have been built with seven or eight blades in an attempt to retain the excavated material in the proximity of the intake. The author was involved in the development of an 8-blade design, Fig. 13-8, which was intended to increase the percent solids going to the pump. It ran smoothly but worked little better than the 5-blade spider it replaced, and once again it was found that the problem was too low a suction velocity. It was during this effort that the idea of the Dredge Laws was conceived as a way of explaining the various dredging phenomena.

Rake Angle

One of the more important design elements of the basket cutter is the rake angle, Fig. 13-9. If the angle is too low, inadequate material will be excavated and production will be low. If the angle is too high, excessive horsepower will be utilized and material rejected. A practical range is 25° to 30°.

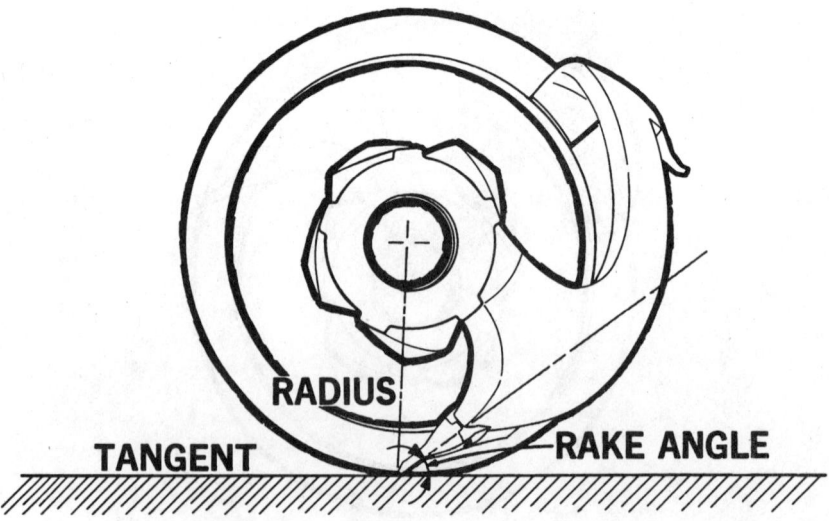

Fig. 13-9. Cutter rake angle. Courtesy: World Dredging Conference.

THE CUTTER

Outside Diameter/Length Ratio

An important ratio is cutter outside diameter to length. A proven reasonable figure is 0.67. If the ratio is as high as 1.0, the crown, where much digging occurs, is at such a distance from the suction inlet as to diminish the transport efficiency.

Peripheral Speed

Little or no laboratory data is available on the correct cutting edge speed for various materials since virgin material cannot be transferred to the laboratory. The conventional wisdom and experience recognizes that a variable speed drive is desirable in order to optimize speed for varying materials. For granular, freeflowing materials, almost any speed succeeds. For hard coral or limestone, a high-speed milling action is recommended, coupled with pinned, hardened teeth. For clay, moderate to high speeds as a function of the clay's consistency is recommended, with either a plain or serrated edge, or the pinned tooth cutter. On blasted rock, a slow speed is dictated to avoid repelling the particles. With the variable speed, the operator can experiment on various materials to optimize his cutter operation.

Most cutters have a maximum peripheral speed varying between 300 and 600 feet per minute. The speed variation should be capable of at least a 50 percent reduction, normally with constant torque, and preferably, but not essentially, a smooth, stepless reduction. If the drive is an alternating electric current, it will probably have a stepped reduction, which limits its flexibility, but does not disqualify the drive. For hard materials, a top speed of 600 feet per minute maximum is recommended; for softer, more normal material, a top speed of 400 feet per minute, with both having speed reduction capability to perhaps 200 feet per minute in order to minimize wear, power, and dispersion of bottom materials when the higher speed is unneeded.

Horsepower vs. Torque vs. Cutting Force

Cutter horsepower can be misleading when quoted as a simple number. On most bottom materials, torque or cutting force is the key to successful excavation, not horsepower alone.

$$HP = \frac{\text{torque} \times \text{RPM}}{5{,}250} \qquad [Equation\ 13\text{-}1]$$

Torque = cutting force × cutter radius   [Equation 13-2]

For a given 3-foot outside diameter cutter, assume there is a choice between two 100-horsepower drives, one with a full-load speed of 50 RPM, and the other 25 RPM. The 50 RPM unit would very likely be cheaper and thus attractive to the buyer, but does it have the same capability as the 25 RPM unit?

In Equation 13-1 above, drop the constant and rearrange to:

$$\text{Torque} \sim \frac{\text{hp}}{\text{RPM}}$$

$$\text{Torque} \sim \frac{100}{25} = 4 \text{ for 25 RPM drive}$$

$$\text{Torque} \sim \frac{100}{50} = 2 \text{ for 50 RPM drive}$$

Equation 13-2 rearranges to:

$$\text{Cutting force} \sim \frac{\text{torque}}{\text{radius}}$$

$$\text{Cutting force} \sim \frac{4}{1.5} = 2.67 \text{ for 25 RPM drive}$$

$$\text{Cutting force} \sim \frac{2}{1.5} = 1.33 \text{ for 50 RPM drive}$$

Therefore, it can be seen that the 50 RPM drive provides only one half the torque and cutting force of the 25 RPM unit.

Cutting force alone is much more indicative of cutter capability than horsepower, but needs to be taken one more step to be definitive. Total cutting force would be sufficiently definitive if all cutters were the same size and geometry, but they are not. Therefore, if one cutter were twice as long as the other, but had the same total cutting force, its force per linear inch of cutter length (pounds/inch) would be only one half as high as the shorter cutter. The operator needs to know pounds/inch of cutter length for a true comparison of cutter options. Successful cutters have varied from 250 pounds/inch to over 2,500 pounds/inch. The requirement is, of course, a function of the material to be dug.

Cutter Drives

The cutter has drive options similar to those of the submerged pump. Variable speed electric drives are impractical to submerge, because of the need to dissipate their heat of inefficiency. Therefore, it is necessary that they use a line shaft. Electrical drives have an advantage over hydraulic in that before stalling, their pullout torque rises dramatically, providing a brief but significant increase in cutting force.

The submerged hydraulic drive has many advantages. It is, undoubtedly, the lowest cost variable speed drive and is relatively simple to submerge. Its speed is easy to change by a simple adjustment to the hydraulic supply pump piston travel. Its torque potential is a constant regardless of speed, providing on demand a constant cutting force which allows a constant relationship to the swing winch line pull. Both electrical and hydraulic cutter drives have their advocates, but hydraulic drives continue to gain on their electrical counterparts in new designs.

Horsepower Requirements

Perhaps the most controversial aspect of cutter drives is their horsepower requirement. One chief executive officer of a major dredging company was heard to remark that there was never a cutter with sufficient horsepower. From the viewpoint of being able to overcome any obstacle and afford the maximum feed to the pump at all times, this sounds like a reasonable statement. However, it is somewhat equivalent to stating that there was never an automobile with sufficient horsepower. It is entirely possible that in a tight passing situation, one might wish he had twice the horsepower on even the most powerful automobile, but the cost of such power would be prohibitive. The same is true of the cutter horsepower on a dredge because the cutting force affects the winches, the spuds, the ladder, and even the hull size. Since the dredge is an economic tool, it is not reasonable to pay a great deal of money for a powerful drive whose full capacity is utilized 1 percent of the time while the cost of its size inefficiency continues unabated 100 percent of the time.

The operator can arrive at a reasonable and economic horsepower for the cutter by working with the peripheral speed of 400 to 600 feet per minute and a unit cutting force of 250 to 2,500 pounds/inch, varying as a function of the material to be dug. In the author's experience, only extraordinary conditions justify exceeding the 2,500 pounds/inch figure.

Velocity and unit cutting force are, of course, a function of cutter outside diameter. A reasonable ratio of cutter outside diameter to suction line inside diameter is 3:1. While this ratio can vary, 3:1 is economical and allows for the reasonable arrangement of a "clown's mouth" suction inlet.

Cutter Calculations

Having determined the size of the suction line from the capacity requirements in Chapter 11, the cutter size and horsepower can be calculated from the following ratios using the above information.

Assume an 18-inch suction and a 3:1 ratio of cutter diameter to suction diameter, then:

$$\text{Cutter diameter} = 3 \times 18 = 54 \text{ inches}$$

With cutter length equalling 0.67 of its diameter:

$$\text{Cutter length} = 0.67 \times 54 = 36 \text{ inches}$$

To calculate the mean diameter of the cutter, assume a 15° face angle:

$$\text{Tan } 15° = .268 \times 36 = 9.6 \text{ inches}$$
$$\text{Diameter at crown} = 54 \text{ inches} - 2(9.6) = 35$$
$$\text{(Say, 36 inches)}$$
$$\text{Mean diameter of cutter} = \frac{54 + 36}{2} = 45 \text{ inches}$$

Assume 500 pounds/inch unit cutting force:

$$\text{Total cutting force} = 500 \times 36 = 18,000 \text{ pounds}$$
$$\text{Torque} = R \times F = \frac{45 \times 18,000}{2 \times 12} = 33,750 \text{ foot-pounds}$$

For hard materials, assume 600 feet/minute peripheral speed, and a pinned tooth cutter:

$$\text{RPM} = \frac{600 \times 12}{\pi \times 45} = 50.93$$

$$\text{HP} = \frac{\text{torque} \times \text{RPM}}{5,250} = \frac{33,750 \times 50.93}{5,250} = 327.4$$

THE CUTTER 125

For softer, more normal materials, use 400 feet/minute peripheral speed and a plain, serrated, or pinned tooth cutter.

$$\text{RPM} = \frac{400 \times 12}{\pi \times 45} = 33.95$$

$$\text{HP} = \frac{33{,}750 \times 33.95}{5{,}250} = 218.25$$

The latter speed and horsepower are more normal for the industry which is not accustomed to thinking in terms of rock being cut by a moderately sized dredge. In many cases, it is more appropriate to blast, but coral, soft limestone, or incipient rock have been dug with no more power than that calculated above for hard materials. The economics of rock dredging are chancy, and should be considered carefully for each project.

The practical operator will recognize that the dimensions arrived at by the above procedure are approximate, and that he should avail himself of economies offered by available standard cutters and drive components which approximate his calculations.

Cutter Capacity

The cutter functions as an excavator and feeder of the solids to the hydraulic transport system. If the cutter is unable to feed the system at the calculated transport rate, the dredge capacity must be down-rated.

Cutter capability varies broadly between dredges. Even where two cutters have the same HP, the cutting force of one can be twice that of the other. To compare cutters, it is necessary to reduce the analysis to the lowest common denominator, i.e., cutting force expressed in pounds/linear inch of projected blade length. Then, by plotting the Standard Penetration Test (SPT) blow count (the dredge industry's traditional indication of cutting difficulty) against the cutting force in pounds/linear inch, and against observed empirical production rate in cubic yards per hour, we can supply the estimator with a guide for predicting cutter limitations on the production of a dredge.

The dredge cutter capacity chart is shown as Fig. 13-10. Note the abscissa between 10 and 100 is the SPT blow count, and the ordinate is cu yd/hr, plotted against various cutting forces ranging from 250–3,000 pounds/linear inch. Using the chart requires no multiplier; rather, if the cu yd/hr of the cutter equals or exceeds the hydraulic transport capability of the dredge, no adjustment is required. If the cutter capability is less, then the dredge capability becomes that of the cutter.

As examples of the use of the cutter plot, note that the capacity of a 10-foot diameter, 250 pound/linear inch cutter on 60 blow count materials is $(10)^2 \times 1.05 = 105$ cu yd/hr; a 500 pound/linear inch cutter would achieve $(10)^2 \times 8 = 800$ cu yd/hr; and a 1,000 pound/linear inch cutter would excavate $(10)^2 \times 100 = 10,000$ cu yd/hr.

It is commonly acknowledged in the industry that there is an advantage of electrical cutter drives over hydraulic drives. This advantage is derived from the "pull-out" or stalling torque characteristics of the electric motor. As the resistance of the soil increases beyond the cutter drive's full load torque, the drive slows down, increasing the amperage and torque substantially before stalling. The stalling torque may be 4–6 times the full load torque, so this temporary torque (obviously it cannot be maintained for a protracted period) adds an estimated 50 percent more effective cutting force than is available with hydraulic power. The HP formulas, which reflect the lower electrical HP requirement for a given cutter service, are shown below.

The HP required for the 1,000 pound/linear inch cutter at 20 RPM where F equals cutting force in pounds/linear inch would be:

Electric HP = $FD^2N/1,970 = 1,000 \times (10)^2 \times 20/1,970 = 1,015$

Hydraulic HP = $FD^2N/1,313 = 1,000 \times (10)^2 \times 20/1,313 = 1,523$

The cutter chart is included with the caveat that the data is the best currently available, but *is not sufficient to constitute proof of the widely extrapolated curves as shown.* There are many shortcomings of the somewhat crude SPT procedure, one of which is its lack of linearity. At times it can make a firmly packed, low porosity sand give the impression of incipient rock; upon excavation, however, such sand disintegrates readily and transports freely if not cemented.

Many geotechnical engineers consider the SPT of dubious value above 100 blow count. The dredgeman needful of the maximum available soil data will extrapolate "refusal" blow count of 75 blows for 3" penetration to an SPT result of 300. These results are nonlinear and questionable, but in the mind of the dredgeman, they are better than blind guesses. There is general agreement that above 100 blow count, a different method of testing is required.

One promising test method sometimes used is the unconfined compressive strength test (UCST). This test is inappropriate for non-cohesive or non-cemented materials since an undisturbed sample cannot be obtained for testing. It is likely that the SPT gives better results on soils with less than 100 blow count, whereas the UCST gives better results on cemented material above 100; modern dredges with powerful cutters have dug lenses of rock with compressive strengths as high as 15,000–20,000 psi. The cutter chart attached has been developed using field data and could prove helpful to the dredge estimator, but should be used with caution.

A clear correlation between SPT and UCST in the range of 10–300 SPT is not clearly established; however the chart allows the use of either SPT or UCST data, as available.

It is possible that a combination of the SPT and UCST will become standard on future projects. Most projects have used the SPT only, necessitating the characterization of some areas as "refusal." This leaves such areas undefined, and the soil data incomplete.

It may prove possible in the future to develop one test for the entire range of soil, such as unconfined shear strength. This may more closely approximate the action of the cutter and give better results. This is an area of research that sorely needs the attention of the dredging industry.

Materials of Construction

It is impractical to harden a one-piece cast steel cutter to a high Brinnell value for better wear since it would fail in shock. Likewise, there is a limit to the hardness and carbon content of cutting edges welded to the softer base blade since the weld would fail if the material were too hard. An advantageous arrangement is the pinned tooth cutter. Here, the teeth which carry the brunt of the excavation can be 350 to 500 Brinnell, while the base blade can be a casting of perhaps 150 Brinnell. A further modification can provide a replaceable cutting edge, possibly 250 Brinnell, which is tack welded to the base blade. The edge can be plain, serrated, or equipped with adaptors to receive the pinned teeth, providing a

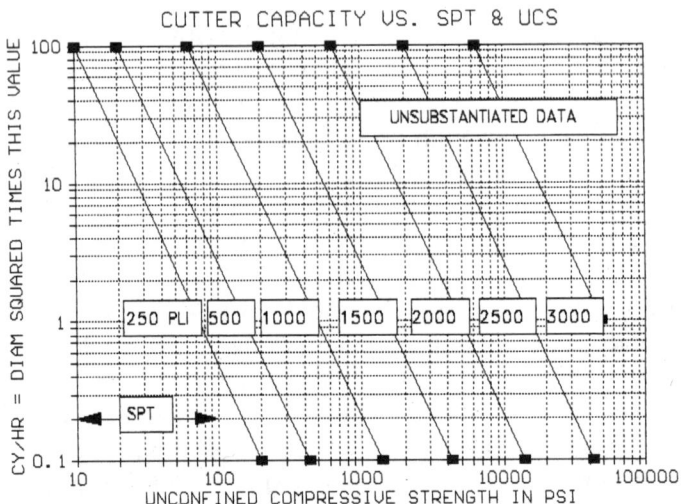

Fig. 13-10. Cutter capacity vs. SPT and UCST.

highly versatile unit. This relatively new configuration is recommended for consideration as a general-purpose cutter. Figs. 13-2 (f) and (g) are examples.

Particle Passage

Surprisingly little coordination between the cutter and the dredge pump regarding particle passage seems to have been attempted in the industry. With the cost in downtime involved in removing oversized particles from the stone box, it would seem some coordination would be justified, but most operators have settled for trash bars welded into the cutter. See Fig. 13-3B. Such trash bars can severely limit the intake of some materials and reduce production; however, if the opening in the cutter would limit the particle size to that which would pass the pump without limiting intake of smaller particles, it would be a boon to the operator. Frequently the sources of the cutter and pump are different manufacturers, and coordination is achieved only by the operator.

THE BUCKET WHEEL

The bucket wheel excavator is a rotating wheel of bottomless buckets mounted on a lateral shaft as shown in Fig. 13-11. The

Fig. 13-11. Rennison Goldfield's 1,340 HP bucketwheel. Courtesy: IHC.

material enters an inner chamber in slurry form, and proceeds to the dredge pump via the suction line.

The bucket wheel was introduced to the dredging industry in the 1970s by an American manufacturer, and since then, has been followed by European and Australian manufacturers. It has numerous advantages which seem to assure it a permanent role in the industry, but it has disadvantages which also assure the role of competitive type cutters.

The conventional basket cutter is, in the vast majority of cases, sold as a separate component. The more complex bucket wheel excavation is normally sold as a complete excavating module, including structure and drive, and the range of size and horsepower offered is more limited than for the basket. No effort will be made in this text to delineate the design aspects of the bucket wheel as was done with the basket shape since it is strongly recommended that only proven proprietary designs of this relatively new excavator be utilized.

Because the excavating element of the bucket wheel is the relatively short length of the bucket projection, each advance of the dredge is much shorter than for the basket cutter. Therefore, the conventional walking spud mechanism is not satisfactory for the bucket wheel since the spud diameter may be equal to the advance, and the spud would fall back in the old bottom hole. The bucket wheel must have a spud carriage arrangement, explained in Chapter 17.

Advantages
The bucket wheel type addresses many of the shortcomings of the basket cutter. The advantages and disadvantages of the bucket wheel over the basket cutter are summarized below.

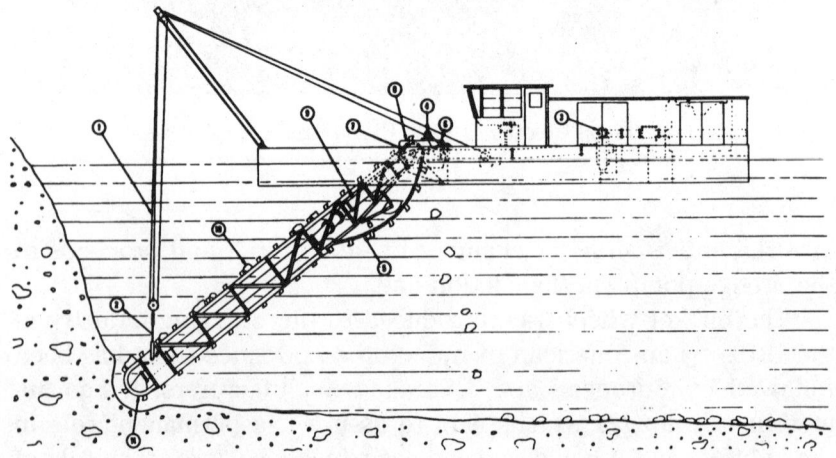

Fig. 13.12. Endless chain cutter. Courtesy: Eagle Iron Works.

(1) Since the bucket wheel cuts on the leading side of the bucket when swinging, it has *bidirectional excavation facility*. Note that the basket cutter digs difficult material effectively only in the undercutting direction. See Fig. 13-6. On the return or overcutting swing, the basket, on firm material, has less capacity, or may be unable to dig at all. The basket cutter also has a tendency to travel or "run away" on the blade edges as the basket rotates. This action can cut the swing wire if it is overtaken, resulting in considerable expense and lost time.

(2) Since the bucket wheel concentrates its horsepower on a smaller length than the basket, a given horsepower can provide as much as three times the unit *cutting force* of the basket.

(3) The bucket wheel has a *positive acquisition* attribute in that it not only force-feeds the material into the suction flow, it also prevents the material from escaping. Unlike the basket cutter which drops material out of its trailing half, sending to the suction pipe only that material which comes under the influence of the high velocity water, the bucket wheel is a more efficient and precise tool.

(4) With its positive acquisition attribute, the bucket wheel can handle *heavy mineral acquisitions*. The basket cutter has never succeeded in the placer mining of gold since it excavates much more material than is transported and the lighter elements tend to rise to the suction inlet and the heavier components tend to drop out to be left behind, enriching the bottom. The bucket wheel could prove to be a successful placer tool, and a viable successor to the expensive bucket ladder dredge.

(5) In the mining of phosphates or other alluvial deposits, the bucket wheel has the advantage of better *depth control* in that it does not react to the solid as a function of digging direction. The basket cutter has a tendency to change digging depth with swing direction. In the undercutting direction, the depth is greater; in the overcutting direction, the depth is less. Furthermore, the bucket approach to the bottom remains at the same efficient angle, whereas the basket loses efficiency as it stands on its nose at greater depths.

Disadvantages

The disadvantages of the bucket wheel relate primarily to cost. Not only is the wheel more expensive than the basket, but the required adjunct of the sliding spud carriage is considerably more expensive than the walking spud arrangement. However, the extra cost of the spud carriage is generally offset by increased dredge efficiency, and many European dredges have installed the spud carriage with basket cutters for the greater facility.

THE ENDLESS CHAIN

The Swintek dredge as shown on Fig. 13-12 is essentially a plain suction dredge which utilizes an endless conveyor chain to scarify the bottom and keep the plain suction clear by transporting oversized particles away from the digging area. It has had broad and successful application in noncohesive deposits of sand and gravel, but lacks the versatility of the basket or bucket wheel in cohesive deposits.

The endless chain, at first glance, may not appear to justify the term cutter. But, because it runs interference for the suction intake, places the intake in close proximity to the material, disintegrates the material to a degree (plus removing oversized particles), the

Fig. 13-13. High speed disc cutter composite.

endless chain performs the functions of a cutter and deserves the title.

THE HIGH SPEED DISC CUTTER

One of the most serious problems of the basket cutter is its inability to deal with tenacious, fibrous vegetation. The peripheral speed of the basket is relatively slow, and there is no opposing anvil to allow the shearing of cellulosic materials which tend to obstruct the flow of soil to the suction inlet. Many a dredging project has suffered production penalties and financial losses because of an inability to deal with vegetation. The disc cutter, equipped with an anvil, shows much success in dealing with vegetation. Without the anvil, it has been successful in sand, clay, and incipient rock.

The disc cutter is patterned after the wood chipper of the paper industry. See Fig. 13-13. It has replaceable knives, protruding from a flat, round disc, which cut the material and force-feed the particles through apertures in the disc where the knives are attached. The stationary chamber on the backside of the disc connects to the suction pipe going to the dredge pump.

Fig. 13-14. 600 HP dual bucketwheel excavator. Courtesy: Ellicott Machine Corp.

Fig. 13-15. Cutter module with cutter removed to show suction mouth. Courtesy: Mobile Pulley & Machine Works.

The cutter head is mounted on a pivot, so that the head can be placed in a vertical position facing the direction of swing. This cutter, like the bucket wheel, requires a spud carriage to provide the precision necessary to properly position the cutter.

Beside dealing successfully with vegetation, the disc uses its speed and inertia to excavate the bottom material. It requires little force to hold it into the cut, since the knives tend to feed the disc into the material. Therefore this type of cutter has been successful in eliminating anchors, winches, and swing wires in favor of swing cylinders (hydraulic) at the stern, mounted on a trailing spud barge, and has significant advantages in shallow areas where anchors are difficult to place.

The main disadvantage of the disc cutter relates to maintenance of the knives which dull quickly on sand, but need to be sharp for vegetation. The knives work surprisingly well on clay, particulating the material for easy transport. The cost of downtime for, and maintenance of, the disc cutter per cubic yard of material will normally run higher than for the basket, particularly when a mixture of vegetation and abrasive materials are encountered.

SUMMARY

Each of the four types of cutters for the pipeline dredge has its advantages and disadvantages. While the basket is still the prevalent cutter, it is recommended that the operator consider each type cutter for every project, in order to make a conscious decision based on the merits of the cutter and its compatibility with the job conditions.

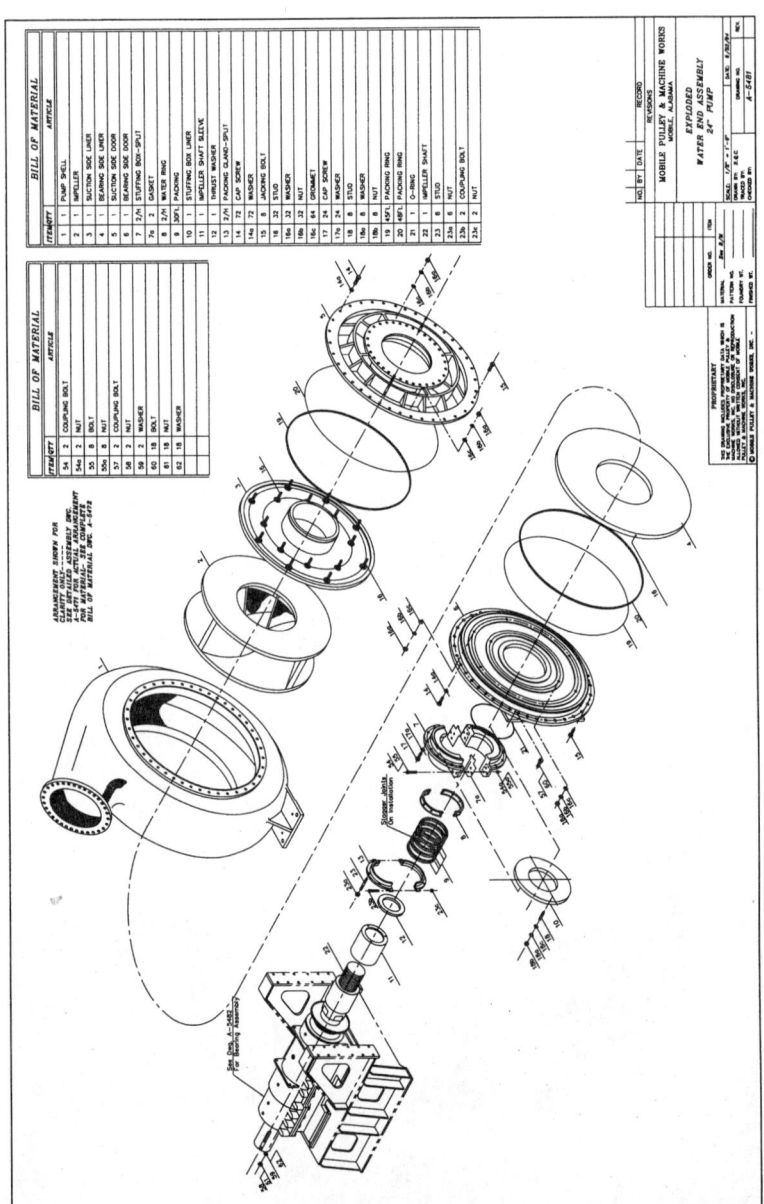

Exploded view of semi-lined pump. Courtesy: Mobile Pulley & Machine Works.

Chapter 14

THE DREDGE PUMP

At the heart of the dredging operation is the centrifugal dredge pump. The centrifugal pump is one of mankind's most useful discoveries; yet it is so commonplace, it is seldom acknowledged as the wondrous machine it is. Consider the following dredge pump characteristics:

1. It induces fluid velocities greater than the speed of its fastest moving component.
2. It creates a vacuum capable of evacuating 90 percent of the atmospheric pressure.
3. It creates discharge pressures between 200 and 300 feet of head.
4. It passes copious quantities of fluid, plus entrained solids.
5. It is surprisingly tolerant of poor design and application, even pumping (poorly) when run backwards.
6. Its overall performance is quite predictable, even though its internal flow pattern is so complex and variable that its details may never be known.
7. It accomplishes all of the above at efficiencies that are the envy of most other machines.

When the dredge pump is efficiently and consistently pumping a good load of solids to the designated disposal area, the operator can be assured he is meeting his economic objectives. The discussion in previous chapters has indicated the effect of project parameters (line size, length, digging depth, etc.) on the performance of well-designed pumps. Unfortunately, all pumps are not well designed. One analyst[5] pointed out that there is a wide and at times disqualifying disparity in the performance of commercially available dredge pumps. The purpose of this chapter is not to detail the mechanical design of the dredge pump, but to provide guidelines to aid the operator in selecting and operating the dredge pump.

PUMP TYPE

The dredge pump should be a single suction, single stage, centrifugal unit with a closed impeller and volute casing. Fig. 14-1 shows an obsolete dredge pump which was designated by its manufacturer as "medium-duty." It differs in some respects from the design of other manufacturers, and is used to provide points of reference for discussion of various dredge pump features.

PARTICLE CLEARANCE

The dredge pump must handle unclassified solid particles of indeterminate size. Regardless of the size designed to pass through it, the pump will eventually encounter a particle which will lodge

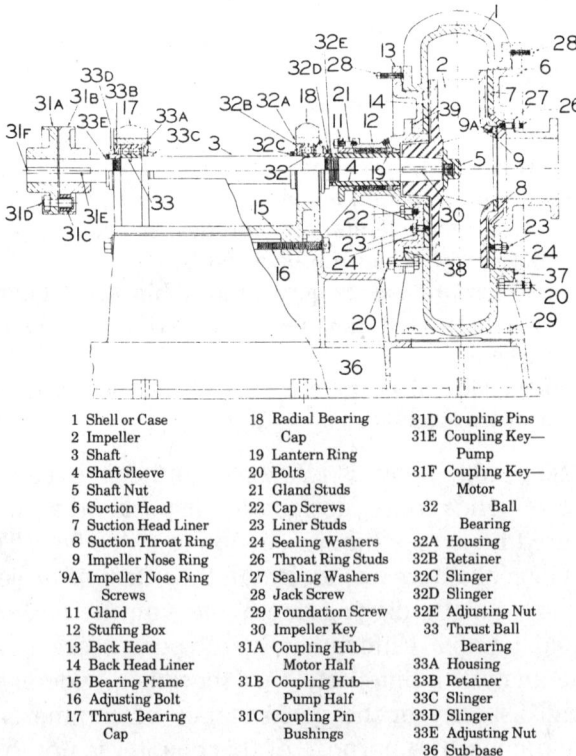

1 Shell or Case	18 Radial Bearing Cap	31D Coupling Pins
2 Impeller	19 Lantern Ring	31E Coupling Key— Pump
3 Shaft	20 Bolts	31F Coupling Key— Motor
4 Shaft Sleeve	21 Gland Studs	
5 Shaft Nut	22 Cap Screws	32 Ball Bearing
6 Suction Head	23 Liner Studs	
7 Suction Head Liner	24 Sealing Washers	32A Housing
8 Suction Throat Ring	26 Throat Ring Studs	32B Retainer
9 Impeller Nose Ring	27 Sealing Washers	32C Slinger
9A Impeller Nose Ring Screws	28 Jack Screw	32D Slinger
11 Gland	29 Foundation Screw	32E Adjusting Nut
12 Stuffing Box	30 Impeller Key	33 Thrust Ball Bearing
13 Back Head	31A Coupling Hub— Motor Half	
14 Back Head Liner		33A Housing
15 Bearing Frame	31B Coupling Hub— Pump Half	33B Retainer
16 Adjusting Bolt		33C Slinger
17 Thrust Bearing Cap	31C Coupling Pin Bushings	33D Slinger
		33E Adjusting Nut
		36 Sub-base

Fig. 14-1 Medium duty dredge pump, now obsolete.

THE DREDGE PUMP

in the suction stone box located immediately ahead of the pump. It is imperative to the economics of the dredge operation that the largest practical particle size be able to pass through the pump so as to minimize the downtime required for cleaning the stone box. This particle size should be a sphere with its diameter the same dimension as the width between the shrouds of the impeller, or a minimum of 60 percent of the discharge diameter of the pump. For example, if the pump has a 20-inch discharge, the width between the shrouds should be a minimum of 12 inches, and the pump impeller vanes should be so arranged as to allow a 12-inch sphere to pass.

Obviously, many pumps do not meet this criterion and, on clean sand, are still able to function satisfactorily. However, this clearance is of paramount importance to the success of a dredge on unclassified material, and the dredges which have foundered on projects with trashy material can bear witness to this fact.

Dredge pump manufacturers, to their credit, have attempted to improve the efficiency of their product by experimenting with the width of the impeller. Some improvement has been noted by narrowing the impeller to more closely emulate the water pump design. Dredge pumps are tested on water, but they are manufactured to pump water *with entrained solids*; therefore, the operator must exercise care to use narrow impellers only where oversized particles will not be encountered. Further, he should recognize that the small improvement in efficiency afforded by the narrow impeller is diminished (and perhaps reversed) by an economic, high SG slurry of *small* particle solids. See Fig. 1-7 regarding the effect of SG and particle size on Che.

FULLY LINED VS. PARTIALLY LINED PUMP

A fully lined pump has the entire inside of the pump enclosure lined with abrasion resisting material. Fig. 14-1 shows a pump that is only partially lined. While it does have a suction throat ring, item 8, it does not have a lining for the full suction throat, which in this case is cast integrally with the front head, item 6. After the abrasive slurry wears out the throat, it is necessary to replace the entire front head with its expensive machining and multiple bolt holes. The front and back heads should not be wearing parts, and in this instance the back head, item 13, is protected by the back head

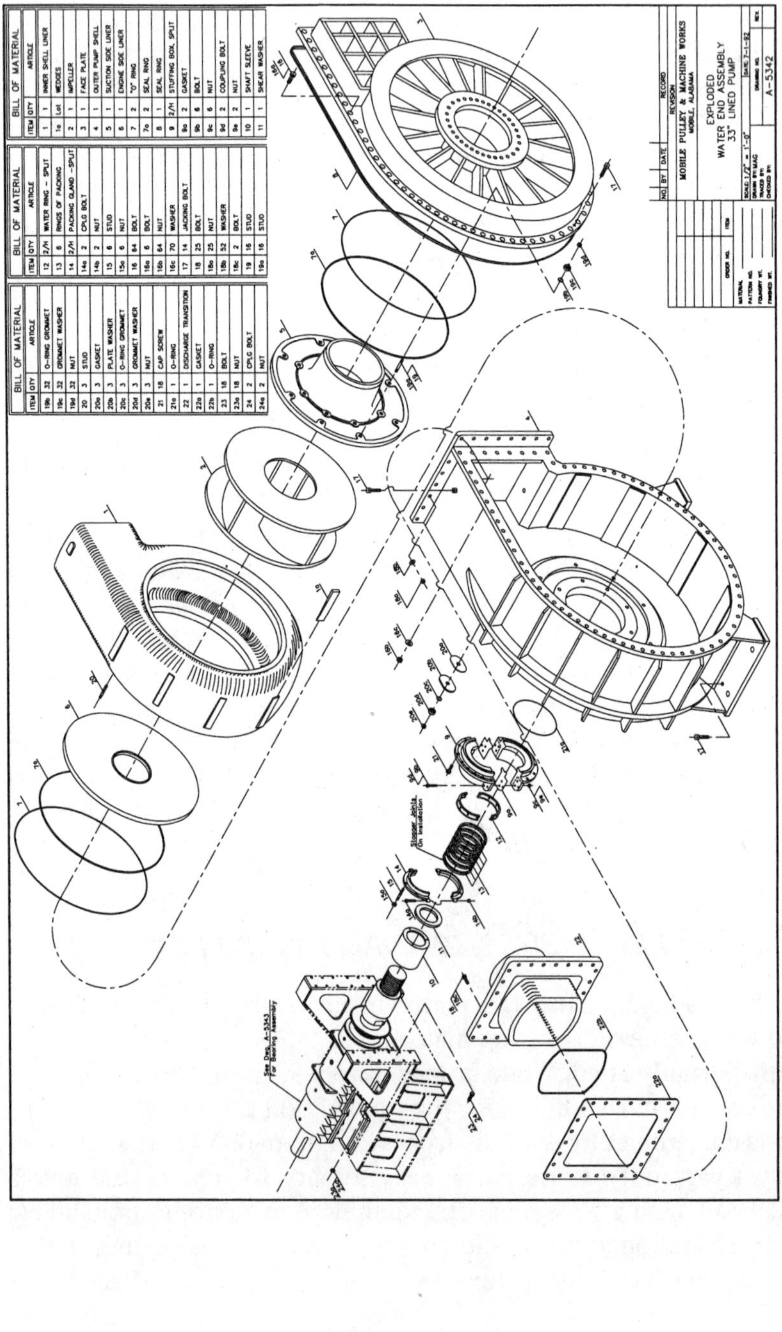

Fig. 14-2. Exploded view of fully lined pump.

liner, item 14. Although the stuffing box housing, item 12, is vulnerable, it could be protected by modifying the design to allow the back head liner, item 14, to extend past the stuffing box.

Most dredge pumps today have full throat liners and front and back head liners. A small minority have shell or casing liners as well. Shell liners increase the size, complexity, and cost of a pump to such an extent that they are seldom used. There are strong advocates for shell liners, however, and there is logic on their side. The author would recommend starting with a pump lined completely *except* for the shell, and letting experience dictate the future course.

IMPELLER

As the pump is the heart of the dredge operation, so the impeller is the heart of the pump itself. The head, capacity, and efficiency of a pump can be radically altered by changing nothing other than the impeller. The author was able to save a dredge pump and a $100,000 drive for a client by replacing a 4-vane impeller with a more effective 3-vane unit that increased significantly its head, capacity, and efficiency. This not only increased production by 40 percent, but the larger particle clearance essentially eliminated the downtime of the pump for cleaning out trash. In this extremely trashy operation, the client had been forced to operate for decades with a large disintegrator (root hog) in the suction line. After the 3-vane impeller was installed, the root hog was scrapped.

Item 2 in Fig. 14-1 is the impeller. It has a curved entrance, which increases its cost, as well as the complexity of its adjacent parts, such as the throat ring, item 9. While there are many who favor this curved entrance, the best performing dredge pumps, with which the author is familiar, have a perfectly straight suction shroud. It is debatable that the increased cost and complexity of the curved entrance is justifiable.

Note also that the front shroud of the impeller in Fig. 14-1 has a larger diameter than the back shroud. This minimizes thrust by minimizing the back shroud area. Sometimes the diameter of the shroud is sized to minimize wear of the adjacent part which is accentuated at its outer diameter. The turbulence and wear at the discharge of the impeller into the casing are best taken by the component part which is longer wearing and easier to replace.

Note that on both shrouds in Fig. 14-1, the wear will be on the expensive casing. If the front shroud had been decreased to the diameter of the back shroud, the wear would have been on the front head liner, item 7, generally a hard, longer wearing material than the casing. Even though they are of the same material, it would still be economical to take the wear on the small, simple, less expensive head liner.

Impellers should always be of the closed type, i.e., have front and back shrouds. The strength afforded to the vanes by the shrouds is essential when the brutal forces of boulders and stumps are encountered.

The impeller illustrated in Fig. 14-1 is keyed to the tapered pump shaft and held in position by a nut on the reduced, threaded end of the shaft. A more common design is to thread the impeller hub and mount it on the male thread of essentially the full diameter shaft against a collar. This latter design facilitates removal by blocking the impeller and rotating the shaft. It also allows the inner hub of the impeller which is exposed to the slurry to be smoothly cast of the impeller material, with no necessity for removing a worn nut.

STUFFING BOX

The stuffing box, item 12, should be replaceable without the necessity for replacing the back head. It should be equipped with at least four rings of packing and a lantern ring for admitting water for cooling and lubrication.

Although it is not yet common, one manufacturer has mounted the stuffing box to the back head through heavy, yielding rubber. This has resulted in an extraordinary long life for the packing, since the nemesis of packing is cavitational vibration which pounds it out. The rubber allows the stuffing box to float with the shaft, greatly reducing the stresses on the packing. Many Europeans have replaced packing with seals successfully, but the simple, packing gland is still prevalent in the United States.

SHAFT AND BEARINGS

The shaft should be generously designed to withstand the buffeting of heavy dredge service. It should be protected in the stuffing box area by a nonrusting, replaceable wear sleeve.

The bearings should be adequately spaced and sized to accommodate the overhung load of the impeller. The bearing adjacent to the stuffing box normally carries a radial load only. This bearing should be carefully sealed and protected by water slingers to prevent water entry from the stuffing box. The other, or outboard, bearing normally carries not only a radial load but the thrust load as well. Frequently a separate antifriction thrust bearing is placed alongside the radial bearing within the common housing. On large pumps, it is not uncommon to use flat bearings with thrust shoes, but the average designer uses this excellent Kingsbury or Union type bearing only when mandatory, because of the high cost.

ADJUSTABLE MOUNTING

Fig. 14-1 shows an excellent cast, machined pump base, with an adjusting screw, item 16, to position the impeller properly within the casing upon assembly, and after wear. This adjustment moves the entire shaft and bearings, requiring some tolerance in the drive coupling, item 31. Not all pumps have this adjustment, some using washers to position the impeller against a shaft collar. The use of washers makes adjustment difficult, and allows volumetric efficiency of the pump to deteriorate as slurry passes between the face of the suction shroud and the front head liner from the high pressure volute to the low pressure suction eye. This passage must be minimized because the higher the flow, the greater the wear, and the lower the pump efficiency. The adjustable mounting feature is a valuable adjunct to the maintenance of a dredge pump, but unfortunately is often not used in actual practice even when supplied.

An alternate method of minimizing recirculation, preferred by some operators, is to have an adjustable throat piece. As wear occurs on the impeller shroud or on the throat piece itself, the throat piece is jacked toward the impeller to reclose the gap. This does not require the loosening of the bearing house bolts, and runs no danger of shaft misalignment.

WIPER VANES

Wiper vanes, sometimes called external or expeller vanes, are located on the outer surfaces of the impeller shrouds for the purpose of precluding or expelling material from the area between the impeller and the head liners. They are not universally used, but in the author's experience, they are justified and recommended. Some plain shroud impellers (without wiper vanes) have reportedly lugged down and even stalled small pumps by the braking action of material between the stationary liner and the rotating shroud. While large pumps with their more powerful drives are not so apt to stall, logic would indicate that the absence of wiper vanes results in losses and wear which are costly to the operator.

The history of the development of wiper vanes is one of simple, pragmatic change. Dredgemen long ago recognized the problems with the plain shroud as shown in Fig. 14-3(A) and attempted to control casting and matching tolerances to minimize the gap between the shroud and liner. They found this impracticable because of high cost, and the difficulty of controlling casting dimensions. Also, when the inevitable wear occurred and the gap widened, the problem again presented itself.

So these practical dredgemen reasoned that if external vanes were added to the shroud, Fig. 14-3(B), the materials would be pumped out of the shroud-liner gap. Their reasoning proved sound. However, the radial external vane with its square cross-sectional shape resulted in severe cavitational wear, particularly at the high tip speed of the periphery.

Cavitation occurs when a mechanical element is moved through a fluid faster than the fluid can rush in to fill the displaced volume. There is typically a high pressure area ahead of the vane and a low pressure area behind it. Most dredgemen have seen the results of this external vane cavitation which wears grooves in the adjacent liner at the approximate outer diameter of the impeller, and severely wears the impeller shroud immediately behind the external vane. Many impellers have been discarded because of this shroud wear, with useful life remaining in the pumping vanes.

The pragmatic dredgeman next reasoned that the external vane could be shaped so as to minimize the cavitation effect. Therefore, he shaped it to approximate that of the more efficient internal pumping vane, and sloped the trailing side to minimize the cavitation, Fig. 14-3(C). This helped, but still did not solve the problem.

In 1970, the writer reasoned that since the problem is one of cavitation (a low pressure phenomenon), it should be possible to devise an external vane to minimize cavitation by avoiding the low

THE DREDGE PUMP

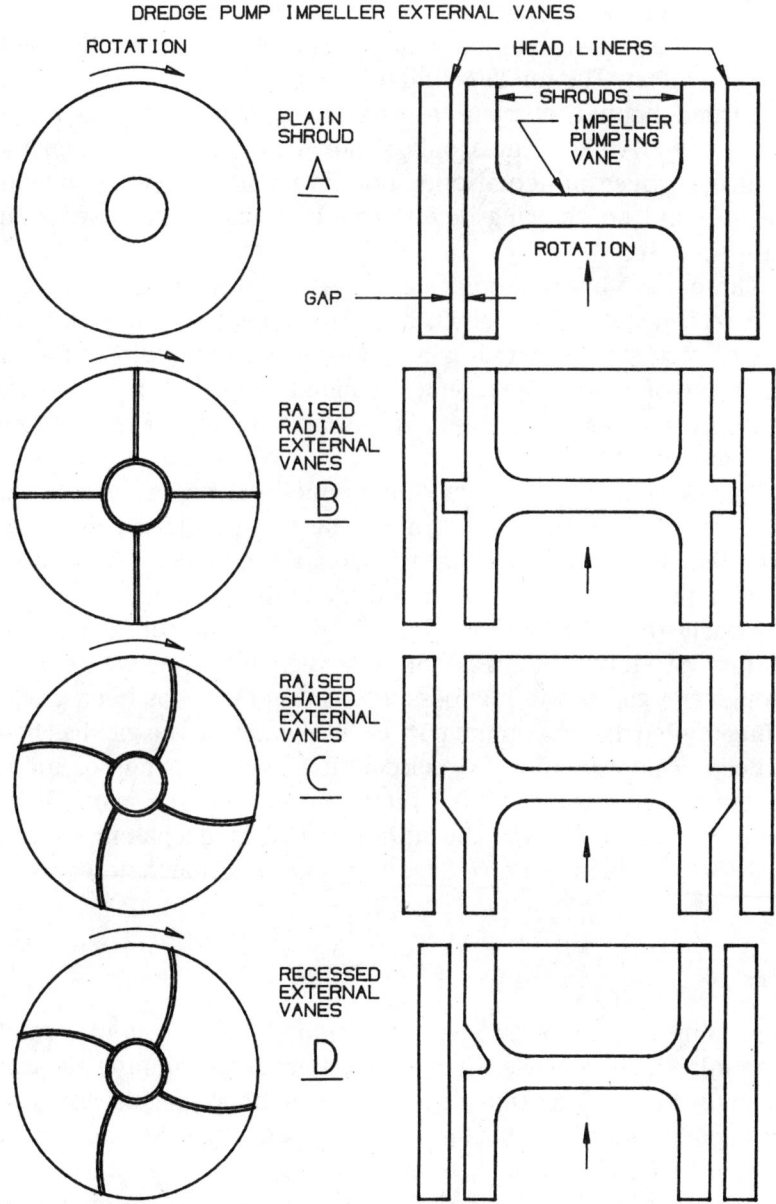

Fig. 14-3. Dredge pump impeller external vane types. Courtesy: Ellicott Machine Corporation.

pressure. With this in mind, the expeller vane shown in Fig. 14-3(D) was devised.

The external impelling surface, although recessed, is the same width as the raised external vane and provides roughly the same pumping effect. The fundamental difference is that there is no trailing surface behind the vane to cause the low pressure cavitation. The recessed vane creates only positive pressure. This principle has been proven in actual operation. Wear has been significantly reduced, and no "lugging down" problem has occurred with the recessed vane.

There are other major advantages of the design also. Note that the recessed vane is opposite the internal pumping vane and, therefore, no structural strength is lost. Moreover, the width of the impeller and, of course, the casing in which it is mounted, is narrower by the height of the external vanes, and therefore less costly for the same capacity.

The mechanical efficiency of the impeller is higher because the losses due to the braking action are largely eliminated. Since the shroud-liner gap (as much as $3/4$ inch for expellers with raised vanes) has been diminished around the entire periphery to the gap previously required between the raised vane and the shroud, the quantity of slurry pumped from the high pressure volute areas through the gap to the low pessure suction area, has been greatly reduced. Also, the maximum particle size entering the gap has been reduced. This reduction in recirculation improves the volumetric efficiency (and output) of the pump, and, of course, partially explains the reduced wear. (The author was granted a patent, October 20, 1970, for his invention of the recessed external vane.)

CASING

The casing of the dredge pump is normally in the form of a volute or spiral, which increases in cross-sectional area from the cutwater to the discharge. Theoretically, this allows for a constant flow from the impeller into the casing around its circumference, and contributes to the efficiency of the pump.

To allow for the passage of large solids, the volute area at the cutwater is normally designed in excess of 50 percent of the discharge pipe area, and increases to 100 percent or more at the dis-

THE DREDGE PUMP   147

charge. This cutwater area is greater for dredge pumps than for water pumps, and is generally conceded to lower efficiency somewhat. There appears to be little doubt that recirculation around the pump increases as the cutwater area increases, but experience has shown that the dredge pump has a wide tolerance for volute area disparities, as demonstrated by the common practice of cutting down the diameter of the impeller to improve horsepower availability on short lines.

Regardless of efficiency, however, the dredge pump must be capable of passing copious solids. The wide impeller, the large volute area at the cutwater, and the 3-vane impeller are all efforts to facilitate the passage of solids. Nevertheless, the modern well-designed dredge pump can still approach a very respectable 80 percent efficiency. With the 10-fold increase in fuel costs of the 1970s, few operators can any longer afford to operate with efficiencies of only 50 to 60 percent.

A note of warning: Not all dredge pump manufacturers have facilities which allow them to gather the data to supply accurate performance curves to the user. Some theoretical curves have proved inaccurate.

EYE SPEED

The impeller eye is the opening in the suction shroud through which the slurry enters. It is normally the size of the suction pipe, although some designs make it larger. The impeller vanes should extend to the eye, and it should be understood that for the purposes of this discussion the eye speed is the speed of the innermost tips of the impeller vanes at the suction shroud.

Eye speed is one of the most critical factors affecting cavitation. If the eye speed is so high that slurry is unable to follow the vane closely, cavities occur which subsequently collapse violently, causing the noisy and vibrant phenomenon called cavitation.

The eye speed should be held to a maximum of about 42 feet per second. There are pumps in existing dredges which can run faster, but almost invariably we find that the top speed range is not used. Therefore, the horsepower in the drive is not fully utilized. If this is true, the gear ratio should be increased so as to keep the eye speed at 42 feet per second or below while demanding more horsepower of the drive by holding the same torque but increasing RPM (horsepower = T × N/5252).

TIP SPEED

Tip speed refers to the velocity of the outer tip of the impeller vane. This is the most significant factor in determining head generated by the pump. Since head varies as the square of the tip speed, and pumping distance is a function of head, the importance of tip speed is apparent. If the tip speed is low, pumping distances may be too low to meet the project requirements. On the other hand, if tip speed is too high, excessive wear of the pump ensues. See Chapter 16 for a discussion on wear.

An analysis of many successful and some unsuccessful dredge pumps discloses a satisfactory range of tip speed would be 95 to 115 feet per second or 5,700 to 6,900 feet per minute which would afford about 175 to 260 feet of head. There may be project conditions which justify tip speeds outside this range, but for general dredging, this range is recommended.

EYE DIAMETER VS. IMPELLER DIAMETER (OD)

It is now obvious that if an impeller is designed for the recommended eye speed of 42 ft/sec and tip speed of 113 ft/sec, the ratio of the impeller OD to eye diameter is:

$$\frac{113}{42} = 2.69$$

This means if we have a 24-inch dredge with a 24-inch suction, a good impeller diameter would be 24 × 2.69 = 65 inches. If the suction were 26 inches, the impeller would be 70 inches. One of the most successful 24-inch pumps in the industry has a 68-inch impeller. A similar success is a 32-inch suction, 30-inch dredge with an 84-inch impeller, a 2.625 ratio. This ratio of eye to impeller outside diameter is not sacrosanct, but it is an excellent guideline. Deviation too far on the low side of the ratio could compromise impeller performance, whereas exceeding the ratio too far on the high side could be costly by creating an unnecessarily large pump.

THE DREDGE PUMP 149

Eye diameter in conjunction with pump RPM is very important to pump cavitation performance. See Fig. 11-1.

HORSEPOWER COEFFICIENT, C_{HP}

The recommended value of C_{HP} is 1.0 for a contract pipeline dredge. This coefficient defines a reasonable HP for any size dredge pump. The higher the coefficient, the farther the dredge can pump; however, higher coefficients require higher HP and can cause faster pump wear. A C_{HP} higher than 1.2 on abrasive material is a dubious application.

DRIVE

A dredge pump can be driven by a steam engine (reciprocating), a steam turbine, a gas turbine, a gasoline engine, a D.C. electric motor, an A.C. electric motor, or a diesel engine.

Steam drives, reciprocating or turbine, were at one time prevalent on dredges. Now, because of their weight, feedwater demands, and other disadvantages, they are obsolescent. The gas turbine, in spite of the concentrated development of recent years, still has a fuel consumption disadvantage which normally disqualifies it. The gasoline engine is almost never used because of the extreme hazard of storing the large quantities of highly flammable fuel on board. Also, it is normally a higher cost fuel than diesel oil.

The D.C. electric motor is an excellent, efficient application for a dredge pump, but its first cost is so high that it is seldom used. Also, direct current is expensive to transmit, and so the generator must be on board or a major rectification of alternating current must be made.

The A.C. electric motor, normally a wound rotor type, is a satisfactory drive, but has an efficiency problem at less than full speed. Roughly, the efficiency of a wound rotor motor is proportional to the percentage of full speed at which it runs, i.e., if the motor is running at 75 percent of full speed, it is also running roughly at 75 percent efficiency. Since the motor runs at full speed only on long lines, the cost of the lower efficiencies is considerable. Also, the generator must be on board, unless the dredge has a constant source of nearby power.

The diesel engine is by far the most common method of driving the modern dredge pump. Whereas the A.C. motor requires a diesel to drive a generator which transmits power through expensive switchgear to the motor which in turn is connected to the pump, the slow speed diesel can connect directly to the pump shaft. More common, however, are medium speed diesels (competitively priced because of mass production) of 900 to 1,800 RPM which drive the pump through a gear reducer.

The marriage of the dredge pump with a diesel engine is an almost ideal union because of their compatible characteristics. First, the diesel is essentially a constant torque machine. This means that the horsepower output is directly proportional to the RPM. In Fig. 14-4, the full load horsepower is 970 at the rated speed of 1,800 RPM. At 1,500 RPM, the horsepower is approximately 800, the ratio of 1,500/1,800 × 970.

Assume that the pump is running at full horsepower of 970 at 1,800 RPM when the specific gravity of the slurry increases from 1.2 to 1.25. At this point the engine would "lug down" in accordance with the lug curve of Fig. 13-3. Since the pump horsepower

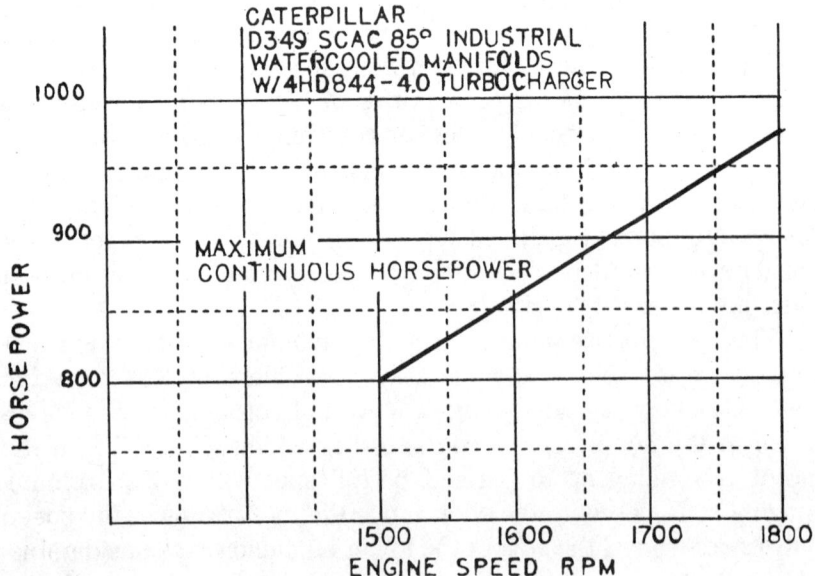

Fig. 14-4. Lug curve, Caterpillar D-349. Courtesy: Caterpillar Tractor Co.

THE DREDGE PUMP   151

varies directly as the specific gravity, the horsepower demand will rise about 4 percent. The engine will put out only its maximum, or rated horsepower, and it must slow down to work within its capabilities. As the horsepower demand of a centrifugal pump on a given system varies as the cube of the RPM, the following relationship exists between RPM and pump horsepower requirement.

Percent Full Speed	Required Horsepower Percent
100	100
99	97
98	94
97	91
96	88
95	86
90	73
85	61
80	51
75	42

It can be seen that in order to reduce the horsepower demand by the required 4 percent, the RPM needs to be reduced slightly over 1 percent. A 2 percent reduction in RPM provides a reduction of 6 percent in horsepower, and a 20 percent reduction in RPM causes a reduction in horsepower of almost 50 percent.

Since a 20 to 25 percent reduction in RPM is well within the operational range of the diesel, and the efficiency loss of the engine at reduced speeds is quite minor (unlike the A.C. motor), it is easy to see why the diesel engine and dredge pump are well mated.

Notwithstanding the above, one of the most frequent problems in the industry is the mismatch found on many existing dredge pumps and their drives. It can occur when the operator decides he needs more pump HP and acquires a bigger prime-mover without changing the pump and/or RPM to utilize the increased HP. The impeller may be too small, or the maximum RPM too slow to absorb the extra power. If he speeds up the pump, he may well encounter cavitation before reaching full engine speed, where full HP is available. In either event, a pump would exist with the inability to utilize the drive HP, an inefficient, uneconomic condition.

The operator should consider using a single competent supplier for both dredge pump and drive to assure a proper match and undivided responsibility; otherwise, procurement of the pump and drive as separate items requires careful design coordination for proper application. This function can be performed by the same

Fig. 14-5. A 27-inch dredge pump. Courtesy: Mobile Pulley & Machine Works.

computer program which calculates dredge rate. Such a program is an invaluable tool for the operator.

TORSIONAL VIBRATION

One of the problems of an internal combustion engine drive is the possibility of torsional vibration. It is not a certain occurrence, but it is a possibility which can be reasonably predicted by conducting a torsional vibration analysis. Many drives have been designed without the analysis and have had no difficulty. On the other hand, some drives have had debilitating vibration requiring a coupling change or some other drive rearrangement.

If a new drive is being purchased, it is recommended that the analysis be conducted, since the engine manufacturer is normally equipped with a computer program to run it. Generally, it is conducted at no charge or at a modest fee as a service to the customer. The computer program requires physical data from the gearbox,

THE DREDGE PUMP

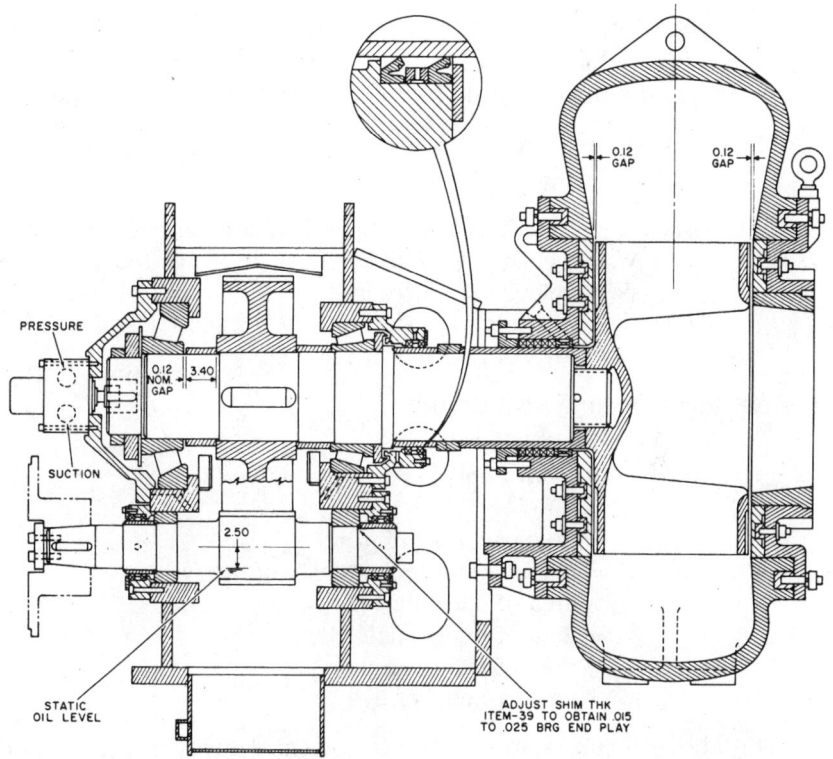

Fig. 14-6. Close-coupled dredge pump. Courtesy: Ellicott Machine Corporation.

couplings, shaft, and impeller which is available from the equipment dimensions. The engine manufacturer can spell out in detail the requirements of the computer program.

THRUST

Pump thrust is that axial force on the impeller which tends to move the impeller toward the suction head. It is primarily caused by the pump-generated pressure on the outside area of the back impeller shroud, which is greater than the area of the front shroud because of the eye opening.

The maximum theoretical thrust can be calculated by assuming that the maximum pump pressure communicates itself over the entire outside area of both shrouds. The shroud forces operate in opposite directions, and the suction shroud pressure is deductible from the engine side. Also, added to the suction side force, and deductible, is the impact force of the slurry stream on the impeller.

$$F = \frac{WV^2A}{g}$$

Where F = force in pounds on impeller toward engine
W = weight in pounds of slurry per cubic foot
V = velocity of slurry in ft/sec
g = acceleration of gravity (32.2 ft/sec^2)
A = area of impeller eye in feet

The net thrust formula would be:

$$\text{Thrust} = 0.433\,h \times SG\,[A_B - A_S) - (A_F - A_E)] - \frac{WV^2A}{g}$$

Where h = maximum head of pump
SG = maximum specific gravity of slurry
A_B = area of back shroud
A_S = area of pump shaft
A_F = area of front shroud
A_E = area of impeller eye

The above formula appears quite logical. However, the author has participated in thrust tests which indicate that the formula may be either inadequate or unnecessarily complicated. For example, the tests indicated that with multiple, effective wiper vanes on the engine side, it is possible to evacuate the back shroud area to the point of achieving a negative thrust. The effect of unusual internal geometry, of eddy currents within the impeller, of solid particles wedging between the shroud and head liner, etc., are so unpredictable as to potentially negate the logical calculation.

To select an appropriate thrust bearing: (1) Use the pump manufacturer's experience wherever possible. Note that if the pump RPM is increased over the manufacturer's experience, the thrust load will increase. (2) In the absence of actual experience, the following simple thrust calculations will generally suffice.

$$T = 0.02\,h\,d^2$$

Where T = thrust in pounds
h = maximum head from pump curve in feet of water
d = diameter of back shroud in inches

THE DREDGE PUMP

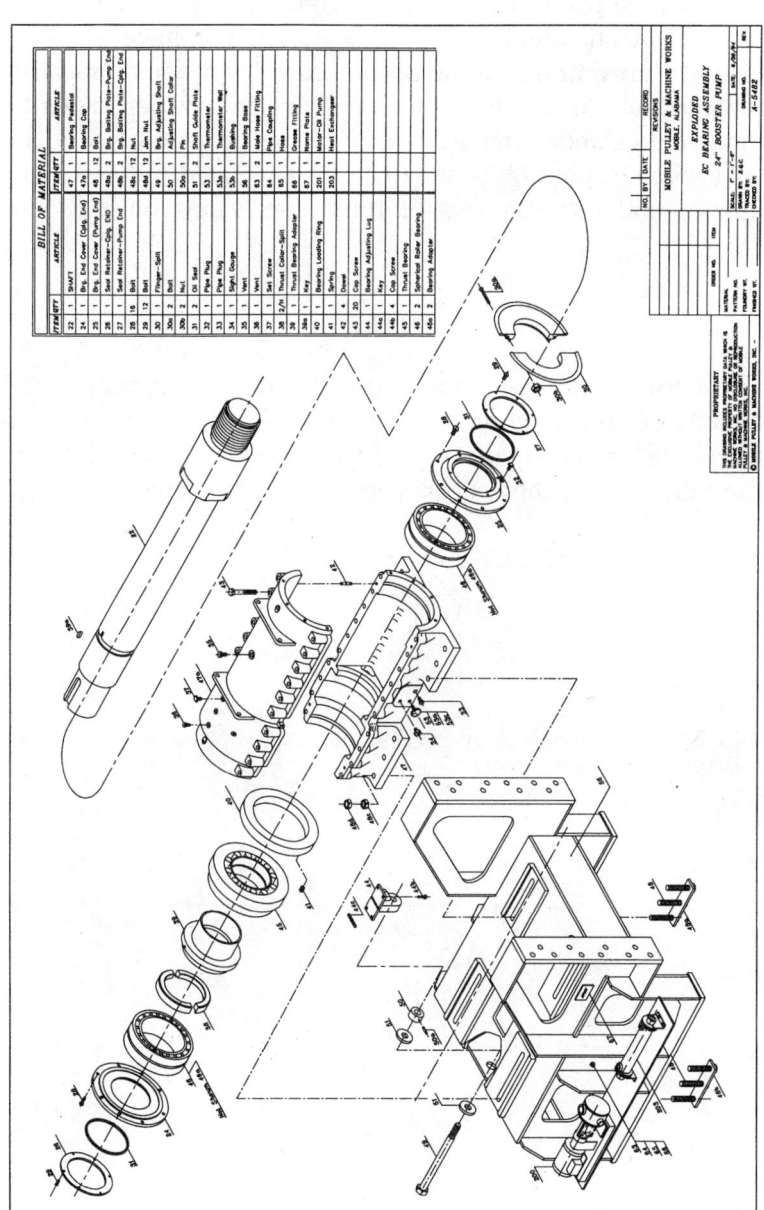

Fig. 14-7. Exploded view of pump bearing assembly. Courtesy: Mobile Pulley & Machine Works.

To provide for thrust, it is recommended that an adequate thrust bearing be used. It is not good practice to cut holes in the back shroud in an attempt to reduce thrust, nor have attempts succeeded in preventing the pump pressure from communicating itself behind the shroud. It is recommended that wiper vanes be used on both shrouds, and if thrust bearing problems do occur, there is a distinct possibility they can be alleviated by increasing the number and effectiveness of the wiper vanes on the *back* shroud.

SUMMARY

When procuring a dredge pump and drive, the dredgeman would be well advised to rely upon test data where available rather than theoretical performance curves, and to evaluate the various offerings carefully against the criteria presented in this chapter.

Dredge ladder structure with lineshafts driving both cutter and ladder pump.
Courtesy: Mobile Pulley & Machine Works.

Chapter 15

LADDER AND BOOSTER PUMPS

LADDER OR SUBMERGED PUMP

Until the 1950s, it was a rare shipping channel that required dredging below the depth of 42 feet. The advent of super cargo vessels, however, has increased channel depth requirements so that today large dredges are commonly specified for digging depths of 70 feet or more.

Operators of conventional dredges digging in excess of 50 feet have discovered to their dismay a sharply reduced output. This has forced some fundamental analysis of the suction system of the hydraulic dredge. The problem, briefly stated, is that there is simply not enough barometric pressure for efficient dredge operation at the greater depths. Dredge Law V addresses this problem as discussed in Chapter 6, and indicates the solution as being the use of the submerged or ladder pump.

The centrifugal dredge pump mounted in the dredge hull is an ingenious machine that, when properly designed, can create a substantial vacuum of 28 feet or more of water; therefore, it can properly be considered a vacuum pump, with the concomitant ability to impel and pass large quantities of soil-laden water while creating substantial pressure or head. This head provides the force necessary to overcome the resistance to flow in the discharge line. However, *the only force to overcome the resistance to flow in the suction line is normal barometric pressure* which forces slurry up the suction line and into the evacuated pump casing.

The vacuum created by a dredge pump is a direct indication of losses and velocity head in the suction line. At the same time, this vacuum is only a rough indication of flow since the vacuum (reduction in barometric pressure) includes all flow-related factors

Fig. 15-1. Ladder pump—900 horsepower electric. Courtesy: Ellicott Machine Corporation.

(entrance loss, friction loss, velocity head) *plus* the specific gravity head, h_{SG}. Specific gravity head represents the force needed to lift the solids content of the slurry from the sea bottom to the dredge pump. Mathematically, it can be expressed as:

$$h_{SG} = \text{digging depth} \times (SG_{slurry} - SG_{water}) \qquad [\textit{Equation 15-1}]$$

Calculations, testing, and actual dredge operations indicate that sand dredging is most efficient at a slurry of 1.5 specific gravity. Using this figure, the parenthetical value in the above expression becomes 0.5, and then:

$$h_{SG} = 0.5 \text{ digging depth}$$

At 30 feet digging depth the specific gravity head value is 15 feet, over half of the 28-foot vacuum the dredge pump can create.

LADDER AND BOOSTER PUMPS

This leaves only 13 feet of the available barometric pressure to provide for velocity head, entrance loss, and friction loss. The premise is defensible that modern dredges should never be built for depths greater than 30 feet without a ladder pump.

At 50 feet digging depth, specific gravity head would be 25 feet if the slurry specific gravity were 1.5. This is obviously an impossible condition since the 3 feet of barometric pressure remaining is insufficient to create the necessary velocity in the suction line to carry the 1.5 specific gravity slurry. Therefore, the operator is forced to reduce his solids content (his payload) in favor of greater velocity. Note that he still uses his maximum vacuum, but there is an enforced redistribution of the 30-foot barometric pressure between specific gravity head and the velocity-related factors. Unfortunately, the operator is not always provided with the necessary instrumentation to know what redistribution he makes. He sorely needs a production meter. If dredge owners clearly understood the importance of this tool, few dredges would be operated today without such a device.

The production meter does not solve the problem of deep digging, however. It merely allows the operator to optimize his operating conditions regardless of depth. The solution to deep digging is a submerged pump, which when properly applied, eliminates for all practical purposes the barometric restriction.

DESIGN REQUIREMENTS FOR LADDER PUMPS

These requirements are as follows:

(1) Head-capacity characteristics coordinated with those of the hull dredge pump in order to achieve the hydraulic transport system characteristics required
(2) Relatively maintenance-free operation and capability of passing particles as large as, or larger than, the dredge pump to minimize downtime
(3) Installation as far down the ladder as possible, but never less than one-third of the distance down the ladder
(4) The minimum head provided by the submerged pump should be one-half of the digging depth plus 15 feet
(5) A constant speed drive, as long as the ladder pump discharges to a variable speed dredge pump; otherwise a variable speed drive

To show the effect of a submerged pump on the production of a dredge, see the calculated production chart shown in Fig. 8-5. Note the production of the conventional dredge at 10 feet digging depth and 50 feet digging depth as plotted against discharge line length. Then note the production of the same dredge, but with a submerged pump, as shown at 50 feet digging depth. It is significant that, with the conventional dredge, the output at 50 feet digging depth approximates one-half of that at 10 feet digging depth. The production of the dredge with submerged pump increases dramatically at the greater digging depths, beyond that of the conventional dredge at the shallow depth of 10 feet. The explanation of this lies in the fact that even at 10 feet digging depth, the specific gravity head requirement is 5 feet and therefore the amount of barometric pressure available for the velocity factors is limited. With the submerged pump, this limitation is practicably removed.

SUCTION JET BOOSTER

The suction jet booster has received considerable attention as attempts have been made to solve the suction problems of a hydraulic dredge. Its success has been mixed, with some applications having been removed as unsuccessful or uneconomic. There is no example known to the author of a submerged centrifugal pump having been installed and subsequently removed. The reasons for the mixed success of the suction jet booster are probably many, including poor design, poor application (less is known of the jet booster's performance than is known of the centrifugal pump's performance), but the greatest shortcoming of the jet booster system is its inherent reduction in the solids-carrying capacity of the dredge pump system.

To demonstrate this, consider that one of the foremost suppliers of suction jet boosters recommends that 40 percent of the total water flow be injected as clear water into the jetting device. This 40 percent of the slurry system does not pick up solids at the suction head, thereby allowing only 60 percent of the total flow to be loaded by solids. Compared to the submerged pump system where 100 percent of the flow is available to pick up and transport solids, there is a significant capacity difference of 67 percent in

Fig. 15-2. A 30-inch electric ladder pump. Courtesy: Ellicott Machine Corp.

favor of the submerged pump system:

$$\frac{100 - 60}{60} = .67$$

When efficiency is considered along with the production factor, the suction jet booster is seldom attractive. An analysis of the circumferential suction jet system efficiency, from the prime mover through the water pump and jet itself, discloses an efficiency of under 20 percent. A comparable analysis of the submerged pump system discloses an overall efficiency considerably higher.

It should not be construed from the above that there is never an economic application for the suction jet booster. In the event it is not feasible to install a submerged pump in a deep digging dredge operation, the suction jet booster can provide relief. A test was conducted on a West Coast dredge which resulted in the conclusion that under deep-digging conditions, the jet suction booster resulted in a 30 percent increase in production over a conventional dredge. It should be noted that the production of a conventional

Fig. 15-3. A 12-inch ladder pump, hydraulic drive. Courtesy: Ellicott Machine Corp.

dredge at 50 feet digging depth is only 50 percent of that at 10 feet digging depth. Therefore, if the 50 feet digging depth production is increased by 30 percent, the dredge capacity is increased to 65 percent of that at 10 feet digging depth. Since the capacity of the same dredge at 60 feet digging depth with submerged pump is approximately 130 percent of the conventional dredge's productivity at 10 feet digging depth, and therefore twice as productive as the jet booster system, one can conclude that the above claim of 67 percent increase is perhaps modest.

NATURAL GAS PROBLEMS

In addition to production advantages, the submerged pump can solve "gas" problems in the suction by never allowing the released gas to expand appreciably. At 70 feet digging depth, 1 cubic foot of gas liberated by the standard dredge expands to 10 cubic feet at the dredge pump. This expansion results from a simple application of Boyle's Law, where $P_1V_1 = P_2V_2$. Experienced dredgemen know the effect on production of such a situation where serious gas cavitation occurs. The properly installed submerged pump keeps the gas under pressure, precluding the debilitating gas expansion. Also, the higher concentration of solids per gallon of slurry significantly reduces the amount of water to be handled and the sedimentation problem of the flow from the disposal area, thus reducing ecological complaints.

LADDER PUMP DRIVES

Successful ladder pump drives of fixed and variable speed have been utilized. The fixed drive units discharge into a variable speed dredge pump which provides the flow adjustment for varying line lengths. If a dredge is equipped with only the ladder pump, i.e., no dredge pump, a variable speed drive is required. (Theoretically, if the discharge line of a dredge is fixed, a competent designer can design fixed speed pump to do the job, but this imposes such permanent restrictions on the system that it is practically never done.)

Ladder pump drives can be of several types, depending on whether the units are submerged or above water; electric (A.C. or D.C.) or hydraulic; and of constant or variable speed.

The above-water drive requires a line shaft with its multiple bearings and requirement for a rugged ladder to maintain alignment. Either electric or hydraulic drive can be used with the line shaft, and either constant or variable drive. If a submerged drive is to be used, it can either be electric or hydraulic as long as its speed is constant. The complexities of a D.C. electric or other form of variable electric drive make it difficult to maintain underwater, while dissipating the heat of inefficiency. The hydraulic drive is superior in underwater application because it is already sealed against its environment and its speed is changed by varying the flow of hydraulic fluid to it. If an underwater, variable drive is required, the hydraulic drive has no peer when cost, weight, and versatility are considered.

Submerged fixed speed A.C. electric drives are quite practical, however. They have been successful as fluid-filled motors, as well as cannister-enclosed, air-cooled motors. In the fluid-filled motor, the oil or water coolant is circulated through the motor and cooled by a heat exchanger where required. In the cannister-enclosed unit, the motor is air cooled in the conventional manner, but a blower is mounted on the motor shaft extension and air is circulated through fins on the inside of the cannister, allowing the heat of

Fig. 15-4. A 14-inch booster pump. Courtesy: Ellicott Machine Corp.

inefficiency to be dissipated through the cannister shell to the ambient water.

Underwater drives can be problematical. Therefore, the operator should not be in a hurry to be the first to utilize a new unit without a thorough analysis or complete confidence in the supplier.

BOOSTER PUMPS VS. TRANSPORT DISTANCE

The booster pump is a supplement to the dredge pump, increasing the distance the slurry can be pumped. Dredge Law VI in Chapter 7 establishes that pumping distance is directly proportional to horsepower available on the pumping system. Therefore, the booster pump is brought to bear when the output of the dredge pump is so reduced by long line conditions, that the cost of a booster pump is justifiable by the increased productivity it achieves.

COORDINATION WITH DREDGE PUMP

The question is frequently asked, "Should the booster pump be identical to the dredge pump, and with the same drive?" While production calculations and operational control are made easier if the pump and drive are identical to the dredge pump, neither is essential. For example, a 24-inch dredge has pumped into an 18-inch booster, with reasonable success. Admittedly, this disparity in size stretches the point, but the point is simple. If the flow rate of the boosted system falls on the head-capacity curve of the booster pump where there is net head or energy added to the slurry, the capacity and distance of the pumping system will be enhanced. The efficiency of the 24- to 18-inch system is suspect, since one or both pumps will necessarily be operating far from their best efficiency points.

WATER HAMMER

If the dredge were 18 inches and the booster 24 inches, it could be made to work, but a different problem presents itself. One of the serious problems of a boosted system is water hammer or as dredgeman refer to it "ram-off." This can be more serious in a multiple pump system than with the single dredge pump because there are two pumps where cavitation can occur, particularly with the booster pump having greater capacity than the dredge pump. Also, with the greater velocity potential of two pumps, the resultant ram-off can reach more serious proportions since the maximum pressure rise above the normal system pressure is a multiple of the normal system velocity.

$$h_{max} = \frac{V_{sw}V_d}{g} \qquad [Equation\ 15\text{-}2]$$

Where h_{max} = maximum water hammer pressure increase
V_{sw} = velocity of sound in water
V_d = normal system velocity in discharge pipe
g = acceleration of gravity

Note the similarity between the above expression and velocity head. If we use the approximate value of 4,800 feet per second for sound in fresh water (about 5,000 ft/sec in seawater) and a normal system velocity of 18 feet per second then:

$$h_{max} = \frac{4{,}800 \times 18}{32.2} = 2{,}683 \text{ feet or } 1{,}162 \text{ psi,}$$

a considerable figure.

Equation 15-2 reflects the instantaneous closure of a valve in the system, and therefore represents a magnitude of pressure that is higher than that normally seen in the dredge system. The most common ram-off in a dredge system is due to cavitation when the noncompressible water suddenly loses its impelling force, slows down, tries to reverse direction, and sets up reflected pressure waves acting in both directions. Since both ends of the system are ostensibly open, the pressure waves are not always destructive. However, many a dredgeman can document that when the conditions are right, elbows can be ruptured, pumps damaged, and pipe supports distorted.

One of the more dangerous elements in a dredge system can be the flap valve, a useful device in helping to prime the dredge pump. If the dredge is pumping against a high terminal elevation and the pump loses its prime or is otherwise shut down, the slurry tries to flow back to the dredge. When this occurs, the flap valve

LADDER AND BOOSTER PUMPS

(a simple, swinging check valve) closes abruptly, and water hammer conditions are set up.

LOCATION OF BOOSTER

Perhaps the most common question asked about the booster pump pertains to its location in the system. Assuming the dredge and booster pumps to be of equal horsepower, a simple rule of thumb says that the booster should be located at about 40 percent of the line length. A look at Fig. 15-5, plotting the pressure gradient against line length, discloses the reason.

If we assume both pumps have a capability of 220 feet of head; the friction loss is 5 feet per 100 feet of line; and the dredge pump suction losses including velocity head are 30 feet, then we can draw the pressure gradient for locating the booster pump in various locations.

First, the system without a booster pump would be represented by line A–D. Since the dredge pump has a head capability of 220 feet and 30 feet is used by the suction line, point A is at 190 feet. With a friction loss of 5 feet per 100 feet of line, the production

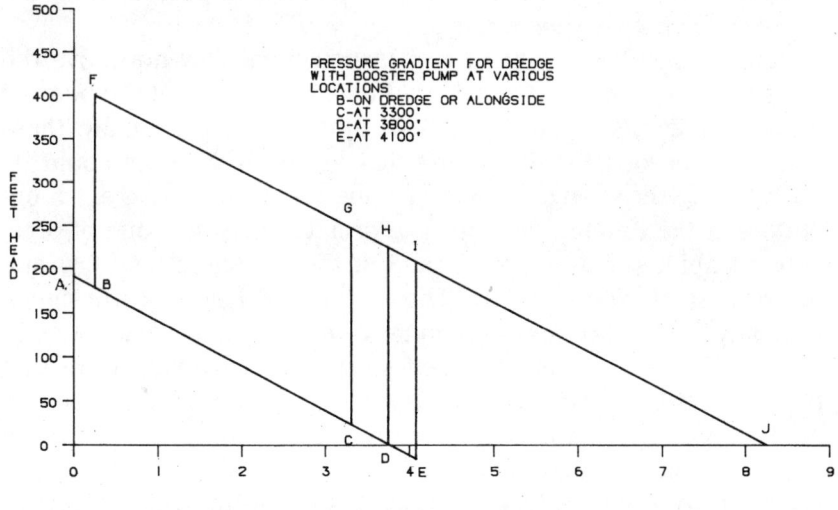

Fig. 15-5. Pressure gradient—dredge with booster pump at various locations.

can be pumped to 190 ÷ 5 × 100 = 3,800 feet, with the pressure gradient falling from 190 feet to zero (atmospheric).

Next, let us assume that we need to pump to 8,200 feet. It may seem logical to locate the booster at 50 percent of the line length at 4,100 feet, point E. We note, however, the pressure drops to 15 feet below atmospheric, and the booster pump is actually drawing a vacuum before it adds its contribution of 220 feet head to raise the pressure to 205 feet at point I. The system then follows the same pressure gradient, 5 feet per 100 feet, to 8,200 feet line length, point J. The negative pressure at point E is hazardous; therefore System AEIJ is not recommended since it could result in serious ram-offs.

If the booster is located at 3,800 feet (46 percent of the line length), System ADHJ is implemented. Here, under ideal conditions, the slurry enters the booster at zero (atmospheric) head. If there were no vacillations in the dredge system, this might be satisfactory. However, a pressure swing of 30 feet of head in the system is relatively common, so that this location is also hazardous.

If the booster is located at 3,300 feet (40 percent of the line length), System ACGJ is effected. Here, the slurry enters the booster at a head of 25 feet, providing a reasonable margin of safety to prevent cavitation. The maximum head realized in the system is 245 feet, easily handled by pump and pipe, so that the booster location is satisfactory.

Since 3,300 feet, point C, is so satisfactory, why not move the booster even closer to the dredge? Indeed, why not put the booster on the dredge itself and implement System ABFJ? Actually, there are good reasons for doing just that since the booster could be operated by the dredge crew, obviating the need for the extra operators at the remote location. The problem becomes one of pressure capability of the booster (and of the discharge line if plastic or worn steel pipe is used). The casing and heads are designed normally for the pressure the pump generates itself. If the booster is operated immediately adjacent to the dredge pump (note that the 200 feet between points A and B is the equivalent line length of the elbows to connect the two pumps), it must withstand the added pressure of the dredge pump, as represented by point F on the pressure gradient where a maximum head of 400 feet is realized. If the booster is not of adequate design, the heads will deflect, opening a space between the pump heads and impeller, resulting

LADDER AND BOOSTER PUMPS 171

Fig. 15-6. Trailer dredge sidearm raised to show draghead and submerged pump. Courtesy: Mobile Pulley & Machine Works.

in excess recirculation from the high pressure volute to the low pressure suction, destroying the volumetric efficiency of the pump.

The cost of increasing the pressure capability of the booster pump heads is relatively modest. The heads are normally non-wearing parts, and the onetime cost of adding sufficient depth to the ribbing to withstand the addition pressure would be minor. The additional cost of ribs on the case, a wearing part, is somewhat more, but represents only a small part of the saving effected by avoiding additional operators for the remote booster location. Having the booster pump adjacent to the dredge pump has advantages that should be considered for all long line work.

SUMMARY

As digging depths and line lengths increase, ladder pumps and booster pumps become more commonplace and justifiable. Proper design, placement, and utilization of these significant tools greatly enhance the productivity and economy of the dredge.

Chapter 16

WEAR IN PUMPS AND PIPELINES

Although the literature of the dredging industry has become more prolific in the last three decades, good information on equipment wear is still scarce. This is unfortunate, because wear is a source of high cost to the dredge operator. Many of the materials pumped are highly abrasive, and wear can not be eliminated; however, there are ways the operator can mitigate wear and costs.

The problem of predicting erosive wear is beset by so many variables that it discourages many dredge operators from attempting a solution. However, because erosive wear has a significant effect on operating time and cost, the operator must understand the principles involved in this phenomenon so that he can discharge his management responsibilities reliably. An understanding of these principles allows him to recognize the qualitative, or directional, effect of variables, which is the first step toward arriving at reasonable quantitative predictions.

Some operators are surprisingly successful at predicting wear, and even on projects where they err, discover in retrospect the variable which caused the error, allowing them to make better subsequent predictions. This chapter will touch upon the major variables involved with erosive wear, and will suggest a simple, mathematical model which can be used as an aid in predicting it and its costs.

LIFE VS. WEAR

The life of dredge components is normally expressed in terms of cubic yards of solids pumped. Therefore, if a pump impeller survived for 1,000,000 cubic yards, its life would be expressed as follows:

$$\text{Life} = \frac{1{,}000{,}000 \text{ yd}^3}{\text{impeller}} \text{ or one million cubic yards per impeller}$$

Wear is the reciprocal of life, and accordingly would be expressed as follows:

$$\text{Wear} = \frac{1 \text{ impeller}}{1{,}000{,}000 \text{ yd}^3} \text{ or one impeller per one million cubic yards}$$

In some instances, it may prove convenient to measure wear in mils (.001 inch) per million cubic yards, or in the metric system, millimeters per million cubic meters of pumped solids. However, in the final analysis, the mils must be translated into life of the dredge component. Since the impeller wears in an irregular fashion, this is difficult if not impossible. But, if experience has indicated that an impeller wears out after passing one million cubic yards, a convenient basis is established for predicting the number of impellers needed for a project, allowing for the inclusion of their cost. Nevertheless, defining the life of an impeller as one million yards without designating the determining variables is as meaningless as stating the capacity of a dredge without indicating material being pumped, line length, digging depth, etc. The soil to be pumped, the job conditions, and the component design affect life (or wear) and all must be considered.

LIFE EQUATION

A mathematical model for the life of dredge components exposed to slurry wear follows:

$$L = \frac{K \times H \times B \times C \times S^{2.5}}{V^3 \times W \times (d_{50})^{0.8} \times A}$$

Where L = life in cubic yards of solids pumped
K = best available prediction of life
H = hydraulic design factor
B = Brinnell hardness of wearing part
C = concentration of slurry (percent solids by in situ volume)
S = size factor (outside diameter of impeller for pump; inside diameter of pipe for pipeline)

V = velocity in ft/sec (of impeller vane tip for pump; of slurry for the pipeline)
W = weight of solids vs. water (SG or density)
d_{50} = median diameter of solid particles
A = angularity or sharpness of solids granules

THE K FACTOR

K is defined as the best available prediction of life of the wearing part. All predictions of the life of a wearing part start with empirical data. There would be no basis for prediction without experience since there is no way to calculate the many imponderables of abrasion. However, once any kind of wear experience is obtained, the mathematical model will allow that experience to be translated into the new job conditions and a prediction for the new job calculated.

HYDRAULIC DESIGN FACTOR, H

The configuration of a pump determines the flow losses through it and affects the wear rate of the pump significantly. One dredgeman, for example, insisted that recessed wiper vanes on the impeller shroud doubled the life of his impeller.

The H factor can only be derived from experience with the specific pump(s) involved. When comparing the future performance of a given pump with its past, the H factor becomes unity and in effect drops from the equation. This would also be true for a geometrically identical pump of a different size, although the size factors would then apply.

THE BRINNELL HARDNESS FACTOR, B

Cornet[6] found that the life of a component varied approximately with its Brinnell hardness. For example, a steel component with a Brinnell of 125 would last one fourth as long as a hard iron component with a Brinnell of 500. While there are other factors such as grain structure, toughness, or friability which affect life, hardness is the most significant characteristic in this regard.

SOLIDS CONCENTRATION, C

Pokrovskaya[7] plots a series of curves from test data that indicate that wear (reciprocal of life) varies with the 0.63 power of concentration up to 15 percent solids by volume. After this point, the wear is constant, presumably because the wearing surface is barraged by the maximum quantity of granules at 15 percent, with higher concentrations resulting in granule on granule contact, shielding the wearing surface from the effect of the greater quantity of solids.

Assuming a dredge operated at 50 percent dredge efficiency, the average percent solids would be 7.5 percent by volume when the wear becomes constant, which can be expressed as C to an exponent of zero (W ~ C^0). Since any value raised to a zero exponent equals one, the value of C for wear becomes unity, and can be ignored in the denominator of the life equation for all dredges which average above 7.5 percent solids, the normal situation. Cornet was unable to detect any effect of concentration on wear, but he indicated that his observed variations in concentration were small. With his efficient, instrumented dredges, the concentrations undoubtedly averaged above 7.5 percent; thus the wear was constant, indicating agreement with Pokrovskaya.

While solids concentration, C, can be ignored in the denominator of the life equation, it is of paramount importance in the numerator. If, for example, C can be increased from an average of 10 to 20 percent, the wear increases not at all, but the cubic yards of material transported doubles, thus doubling the component life in terms of cubic yards.

Note that C appears in the numerator with an exponent of one, indicating life increases directly with concentration.

PUMP SIZE, S

The S factor relates the size of the pump to wear life primarily through an increase in production while other wear factors remain constant.

If a 36-inch impeller pump is compared to a 72-inch impeller pump *with all other proportions increased accordingly*, the flow rate of the larger unit will logically increase at least by the square of the diameter. The reader will recognize that as pipe size in-

WEAR IN PUMPS AND PIPELINES

Fig. 16-1. Worn front head liner. Courtesy: Ellicott Machine Corp.

creases, the flow must increase not only by the square of the diameter, but to the 2.5 power in order to obtain the necessary turbulence to convey the same concentration of solids; and so with the pump. This states then, the production of the pump will normally increase with size to the 2.5 power, while the tip speed of the 72-inch impeller remains the same as the 36-inch unit, creating the same head. Since the circumference of the 72-inch impeller is twice as large and the other dimensions are in proportion, the pump at the same tip speed will achieve a life increase to the 2.5 power of the ratio of impeller diameters.

While no study of wear vs. pump size has been found to support or refute these conclusions, experience in the field does support the longer life of larger pumps. It is recommended that pump size to the 2.5 power be used until experience allows the development of a better figure.

VELOCITY, V

Velocity normally indicates slurry velocity with respect to a stationary surface, such as the pump case or pipeline wall. A dredge pump impeller provides a somewhat different condition, in that it is moving with respect to the slurry. However, it wears as a function of its velocity against the contiguous and stationary head liners, so that its velocity in the life equation is tip speed or peripheral velocity. Impeller wear can be accelerated by high tip speed relative to slurry velocity as shown in Fig. 16-3. However, in the normal operating range, impeller peripheral velocity fits the life equation reasonably well.

Pokrovskaya states that pump wear is a function of velocity to the third power of tip speed, while Cornet, as a result of years of observation of a dredge fleet, states that pump and pipeline wear vary as velocity to the 2.5 to 3.0 power. Pokrovskaya's data was derived from a controlled experimental effort and may be more accurate than the uncontrolled observation of the dredge fleet by Cornet where other factors could have influenced the result. Since Cornet agrees the exponent could be as high as 3, and Pokrovskaya states it is 3, it is recommended that impeller tip speed and slurry velocity cubed be used in the life equation.

WEIGHT OF SOLIDS, W

The kinetic energy of a particle in motion is directly proportional to its mass; therefore, if the density or specific gravity of a transported solid is doubled, it stands to reason that, with other variables unchanged, the solids would impact the wearing component with twice the force. Thus weight of solids is shown in the life equation denominator to the first power, and does not appear in the numerator since life is expressed in volume, not weight.

Most materials pumped by dredges have specific gravity of 2.65 ± .05, and W can normally be ignored. A significant exception is where organic content is high. Generally, where organic content is high, the particle size is small, so there is a double reduction in wear. Under these circumstances, frequently true in maintenance dredging, wear can become a minor problem, an important consideration in submitting a competitive bid.

PARTICLE SIZE, d_{50}

The median diameter particle size of the solids, d_{50}, has been determined to be one of the most significant characteristics in the flow and wear performance of slurry. Dredges do not encounter solids consisting of only one grain size, but the median grain size has proved a reasonably accurate indicator of the flow performance.

Cornet has found in dredging practice that wear varies with the 0.75 to 0.80 power of the median diameter of solid particles. Logically, it might seem that wear would vary with the first power; but in practice, there may be mitigating factors such as the viscosity of the water or the shielding effect of the smaller particles. Admittedly, Cornet's observations of his dredging fleet were not controlled experiments, but his ten years of observations and hundreds of millions of cubic yards dredged are too sound to be refuted on a theoretical basis. Therefore, median diameter of solid particles is shown in the denominator to the 0.8 power. Cornet cites a telling example where the median diameter of a material (gravel) was 8 millimeters and showed a wear 16 times as high as that shown for a material with 0.25 mm (medium sand).

$$\text{Wear} = (8/0.25)^{0.80} = 16$$

ANGULARITY, A

Taylor[8] has taken a microscopic look at soil particles, and has categorized them as: A, well-rounded; B, rounded; C, subrounded; D, subangular; and E, angular. Wellinger[9] has determined that angular grains cause about twice the wear of well-rounded ones. If we then assign a value of one for well-rounded, and two for angular, we obtain the following:

Angularity	Wear Factor
A, well rounded	1.00
B, rounded	1.25
C, subrounded	1.50
D, subangular	1.75
E, angular	2.00

Most dredge people are well aware that sharpness or angularity can have an effect on wear. It is fairly common knowledge that the wear of material in the upper reaches of the Mississippi River is greater than that of the material near New Orleans, which has been worn and tumbled over hundreds of miles. The author was once associated with a job where the material was fine, but unexpectedly angular. The wear experience was twice as high as anticipated by the experienced operator and resulted in a serious financial loss.

APPLICATION OF THE LIFE EQUATION

The foregoing discussion may lead the reader to conclude that the application of the wear-life equation is difficult and complex. Not so; it is extremely simple.

As a first example, let us assume we are using the same pump on which we have previous experience, and the job conditions are the same. In this case, every factor but K (the best available prediction) becomes unity or one, leaving K as the logical result.

Obviously any time we use the same pump, H, B, and S drop out, leaving only K and C in the numerator. If the C (the percent solids of slurry) doubles, life doubles—if it halves, life halves.

As for the denominator, W (weight of solids vs. specific gravity of water) will generally drop out, i.e., assume a value of one, since most dredged materials have a specific gravity of 2.65. If the material is the same as previously pumped, not only W but d_{50} (median diameter of solid particles) and A (angularity of solids granules) drop out; if the line length allows velocity optimization, V (velocity) drops out; but even when these values do change, the equation is a simple division or multiplication problem, or at worst, the raising to a power, a simple matter with the hand-held calculator.

SIMPLIFIED PIPELINE LIFE EQUATION

A simplified life equation, applicable to pipelines *only*, follows:

$$L = \frac{K \times B \times C \times S}{W \times (d_{50})^{0.8} \times A}$$

This is the life equation with H eliminated, the exponent of S reduced from 2.5 to 1, while V^3 had disappeared from the denominator. It is predicated upon good hydraulic practice in the operation of the dredge, i.e., the increase in flow to the 2.5 power of the pipe diameters when a dredge is increased in size. This can be demonstrated by assuming that the pipeline doubles in size; then,

WEAR IN PUMPS AND PIPELINES

from the full life equation:

$$S^{2.5} = (2)^{2.5} = 5.656 \text{ in the numerator}$$

Now we know from Dredge Law III that the velocity in the larger line must increase by the square root of the increase in size:

$$V \sim \sqrt{d_1 \div d_2} \sim \sqrt{2} = 1.414$$

The life equation states that V in the denominator must be raised to the third power:

$$V^3 = (1.414)^3 = 2.828$$

Dividing as in the life equation:

$$\frac{S^{2.5}}{V^3} = \frac{5.656}{2.828} = 2.0,$$

and we obtain the increase in life corresponding to the increase in pipe diameter. This states then, that *the life of the pipe is proportional to the increase in size* if good dredging practice is followed, and everything else is equal. This is not always the case, so the operator is advised to use the simplified form of the life equation with discretion, adjusting appropriately all factors that change.

NORMAL RANGE VALIDITY

The wear life equation only applies within normal dredge operating ranges. It is obvious that the life of a component would be zero cubic yards when the pump is operating under cut-off conditions or at such low velocities that negligible solids are transported. Practical prediction of wear requires practical operation of the dredge.

PRESSURE

Notable by its absence in the life equation is any reference to pressure. This is not an oversight, since pressure, in itself, plays no part in wear. A dredge pipeline may have 10, 100, or 1,000 psi internal pressure, and the wear would vary only as a function of the factors shown in the life equation. Perhaps helpful in understanding this is the concept of a pipeline with 1,000 psi pressure, but with zero

Fig. 16-2. Worn pump case. Courtesy: Ellicott Machine Corp.

velocity. Obviously, no wear occurs, nor is any production achieved. However, when flow begins, the transfer of energy between the slurry particles and the pipe wall commences and wear begins. The pressure can remain at 1,000 or drop to 10 psi and still be of no consequence to wear.

CORROSION

A factor not included in the life equation is corrosion. Corrosion can occur when the pH of the slurry varies to the low (acidic) side of the neutral position of 7, while values somewhat above 7 tend to inhibit corrosion.

Most metals depend upon the development of a protective film for their resistance to corrosion. Surface rust or other oxides inhibit corrosion. In the case of slurry lines, the abrasive solids keep the surface of the pipe or pump free of protective coatings, and set up an ideal regime for rapid erosive-corrosive wear.

Chemicals are so variable and perverse in their attack on metals that no effort is made to include a factor for corrosion in the life of the equation. Therefore, the dredge operator should be doubly wary when bidding on dredging involving industrial waste or tailing pits. When a deviant pH condition is involved, the operator should consider polyurethane pipe, a testing procedure, and an increase in his bid price.

WEAR ZONES

The dredge operator should be aware that operation in the high head, low flow portion of the dredge pump curve is achieved at high wear cost. This is not because of high pressure, but because of low concentration of solids, C, and high impeller tip speed, V, both of which appear in the life equation. Note the four zones of operation (Fig. 16-3) superimposed on a head-capacity curve of a dredge pump. Zone I represents infinite wear and zero life because the flow rate through the pipeline is too low to transport solids. Zone II represents high wear and low life because V is high and C is low. Zone III represents a more normal wear and life where the pump approaches its highest efficiency, V is normal, and C reaches its optimum. Zone IV represents low wear and high life in the pump since C remains constant and production rises directly as flow, while V drops because of low head requirements. However, while

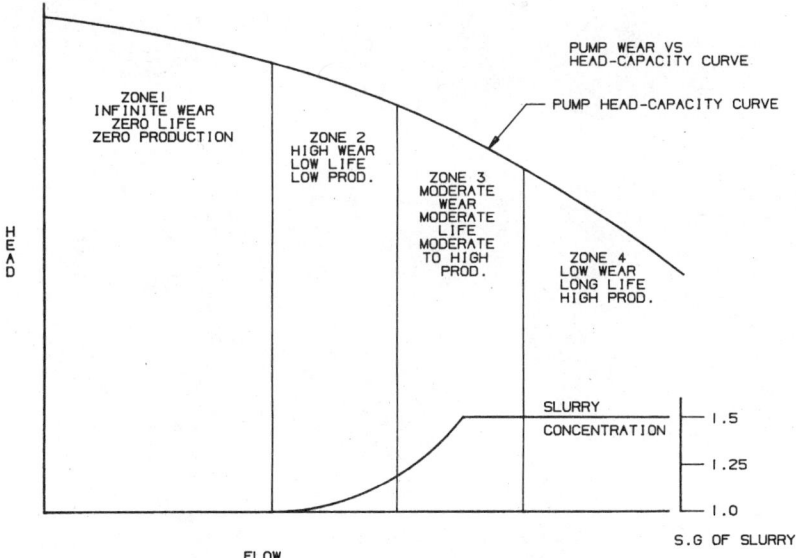

Fig. 16-3. Pump wear vs. head-capacity curve.

this zone of operation is good for the pump, it can be bad for the pipeline. The pipeline equation points out that life increases with C (which reached a maximum in Zone III) and decreases with V^3, i.e., slurry velocity cubed. With an abrasive slurry, high velocities significantly above that required to carry the optimum concentration will exact a high price in pipeline wear.

SUMMARY

Wear is a significant cost, too important for the dredge operator to ignore. Reasonable predictions of wear are possible if the principles are understood and some records are kept. A simple mathematical model (equation) is presented which can guide the operator in controlling and predicting wear and its costs. A simple rule enabling the operator to maximize wear life is to operate with velocities at or slightly above the minimum required to convey the optimum 1.5 specific gravity slurry. A booster pump, if necessary to increase low velocities, can pay off not only in increased production, but in less wear per cubic yard.

Chapter 17

AUXILIARY EQUIPMENT

Foremost among the support equipment on a modern dredge are the winches or hauling gear. Winches can be made up in various forms, but are generally composed of several drums, each for a separate purpose. The normal drums are as follows:

(1) Port swing
(2) Starboard swing
(3) Ladder hoist
(4) Port anchor boom
(5) Starboard anchor boom
(6) Port spud hoist
(7) Starboard spud hoist
(8) Stern mooring (Christmas tree)

FORWARD WINCH

The forward winch is most often a 3-drum winch with a single drive. The drums are for the port and starboard swing, plus the ladder hoist. The single drive necessitates clutches and brakes for each drum, normally more economic than separate drives for each drum. The single, relatively low horsepower drive is possible since the swing drums are never powered simultaneously, although the ladder is occasionally hoisted while swinging. In cases where a side slope must be dug, i.e., where box dredging or terracing on the slope is unacceptable, additional horsepower can be supplied in the drive to allow simultaneous ladder hoisting and swinging. Some dredges have utilized standard hydraulic winches with individual drives for each function, and other combinations have been used successfully. However, the forward winch must meet various rather strict requirements if the dredge is to be an efficient tool.

SWING SPEED

The swing speed must be infinitely variable from zero to full speed. Stepping speeds such as obtained with wound rotor A.C. motors are not satisfactory. Either D.C. electric, eddy current A.C., or hydraulic motors are required to provide the speed control needed for efficient dredging.

Maximum line, or hauling speed, should be in excess of 100 feet per minute in order to conserve time when advancing. A swing speed of 10 to 60 feet per minute is more normal when dredging, and is a function of bank height and difficulty of digging. Full line force may be required when digging, but lesser torque is required when swinging to advance, so that field weakening, e.g., of a D.C. drive would suffice to obtain the higher advance speed.

LINE PULL

The line pull or hauling force of the swing winch is a key attribute of a cutterhead dredge, and is a function of the total cutter force. See Chapter 13. In order for the full cutting force of the cutter to be effective, the line pull of the winch must be sufficient to hold it in the cut, i.e., the line pull must equal the cutting force, plus supply that force needed to overcome the resistance of water, wind, and current to the hull movement. This force is approximately 1.5 to 1.6 times the cutting force. Since the dredge is normally limited to an inclusive swing angle of less than 90°, the swing to each side is less than 45°. The line force required to hold the cutter in the cut is equal to the cutting force normal to the cutter, i.e., 90°. This required force becomes the cutter force divided by the sine of 45°, or 1.414 times the cutter force, when the dredge is at its full swing position. Adding the effect of wind and water, the 1.5 to 1.6 factor becomes apparent.

On some small to moderately sized dredges, the ladder hoist drum is replaced by hydraulic pistons. The pistons have the added capability of "crowding" the cutter down into the soil, whereas with the hoist, only the weight of the ladder can be brought to bear.

Drums should be 15 to 20 times the diameter of the wire to avoid undue stresses on the wire. Grooved drums are occasionally used to improve the spooling of the wire, but are expensive and are normally not essential. Fleet angles (deviation from normal 90° wire approach to drum) should not exceed $1\frac{1}{2}°$. Fleeting sheaves (where the guide sheave slides along its axis) may be necessary as a function of the wire configuration. Some dredges mount the drums on the top section of the ladder in order to have a straight line to the swing sheave, and to avoid the lifting component of the swing wire on the ladder. It is more common, particularly on large dredges, to have the forward winch on the main deck enclosed.

Most winch brakes are regenerative (electric) or counterbalanced (hydraulic). They should be fail-safe, i.e., the brake should require the application of air or hydraulic power to release, and should be set automatically by spring in the event of power failure. The clutch works in an opposite fashion, i.e., engaged by power, and released by spring. A modulated drag should be available on each drum to prevent excessive release and snarling when used on the trailing swing wire. All drums should be freewheeling for ease of wire installation and for setting of anchors.

The forward winch is a complex and important element in the dredge function. Since so many things can go wrong and affect the proper spooling of wire and performance of the winch, most operators like to have the winch in view of the leverman. This is certainly desirable, but many successful dredges bear witness to the fact that it is not mandatory.

ANCHORS

Fig. 17-1 shows various types of anchors available to the dredgeman. Swing anchors must hold if the dredge is to perform. Therefore, the selection of anchor type and size is important. The holding power required must be coordinated with the line pull of the swing wire, which should be 1.5 to 1.6 times the cutting force. In order to avoid anchor slippage, a good rule of thumb would be a holding power of 1.6 to 2.0 times the cutting force to account for transient overloads, particularly with electric cutter drives.

Most anchors in use today are some form of patent anchor such as the Danforth, Stato, or Stevin. The holding power of these anchors can vary from as low as 6 to as high as 30 times their own

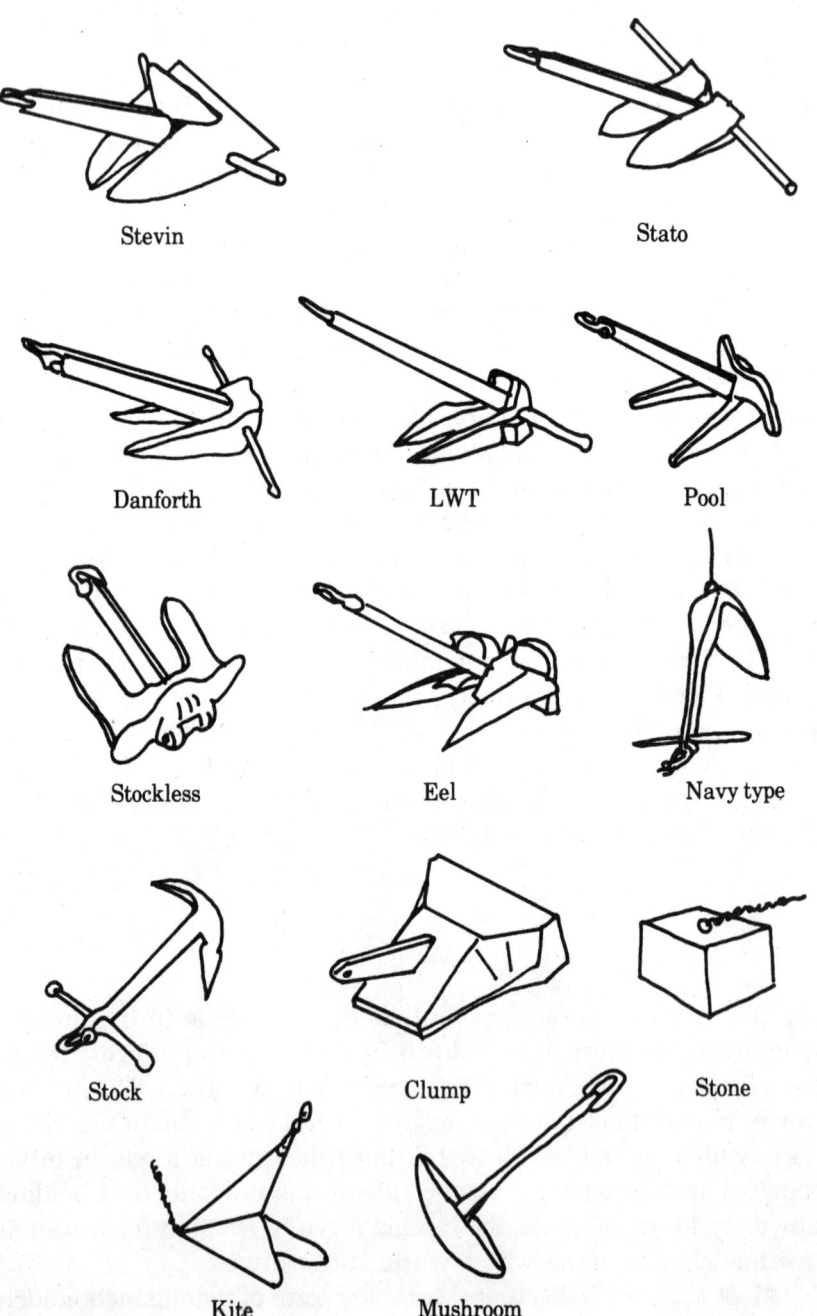

Fig. 17-1. Anchor types. Adapted from Van Den Haak, "Anchors," *Holland Shipbuilding* (October, 1972).

weight as a function of fluke size, fluke angle, palm design, type of bottom being dredged, etc.

Fig. 17-2 shows weights, dimensions, and holding power of various Offdrill anchors. This type of anchor has the desirable facility of wedge inserts to change from a fluke angle of 50° for mud to a more effective 34° angle for sand.

If we were to select an anchor from Fig. 17-2 for the 1,800-pound force of the cutter calculated in Chapter 13, we would multiply by 1.8, the midpoint of the 1.6 to 2.0 recommended range. This calls for a holding power of 32,400 pounds. The 1,000-pound Offdrill anchor would be sufficient since it is rated at 35,500 pounds in sand with a zero degree chain angle. The zero degree chain angle is conservative in many cases for a dredge since a negative angle is often achieved, and the anchor is pulled down rather than up. The anchor rating on mud is only 28,500 pounds, but if mud is being dug, the cutting force and therefore the line pull is probably less than its maximum.

Another important anchor consideration is its breakout requirement when relocation is necessary. The breakout force at the pad eye of various type anchors is estimated at 30 percent or more of the holding power developed.[10] The pad eye is normally on the base of the fluke assembly and, on dredges, is generally attached to the marker buoy. If the anchor has been firmly seated, the anchor barge may have difficulty breaking out the anchor when pulling on the pad eye wire. This breakout force can be reduced in some instances by pulling on the shank shackle as shown in Fig. 17-3. The first illustration shows the force of the tugboat applied through the anchor barge stationed over the anchor. The second shows the swing winch force applied in a similar manner. The Stevin anchor, because of its short fluke ahead of the hinge, claims a significantly reduced breakout force.

ANCHOR BOOMS

Anchor booms are mandatory on a dredge only when a channel is being dug through shallows or other impassible areas, and the anchor barge is unable to deposit the anchors at the appropriate locations. Booms are used by choice by some operators who are convinced that higher dredge efficiencies result. However, the con-

OFFDRILL
High Holding Power Anchor

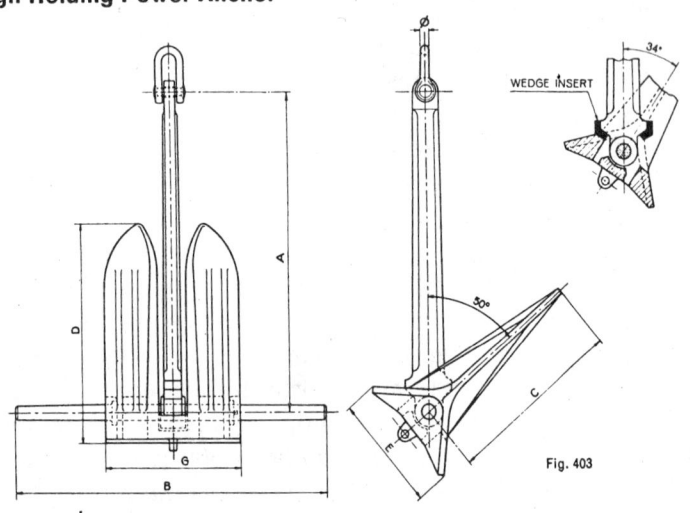

ANCHOR WEIGHT		DIMENSIONS IN INCHES						
Pounds	Kg	A	B	C	D	E	G	Ø SHACKLE
1.000	454	60,83	74,80	36,22	43,11	18,90	25,40	2,00
3.000	1.360	87,80	109,06	52,36	62,21	27,17	36,62	3,00
6.000	2.721	101,18	142,92	60,24	71,65	31,30	42,17	4,00
8.000	3.628	121,86	153,15	72,44	86,22	37,80	50,79	4,00
10.000	4.535	131,30	157,48	78,15	92,92	40,63	54,73	4,50
12.000	5.443	139,37	167,33	83,07	98,62	43,11	58,07	4,50
14.000	6.350	146,85	177,17	87,40	103,94	45,48	61,22	5,00
16.000	7.256	153,54	187,01	91,34	108,66	47,44	63,98	5,00
20.000	9.080	165,36	196,85	98,42	117,13	51,18	68,90	5,50
30.000	13.606	189,37	210,04	112,60	134,06	58,66	78,94	6,50

HOLDING POWER (in lbs.)

ANCHOR WEIGHT		SAND BOTTOM			MUD BOTTOM		
		CHAIN ANGLE DEGREES			CHAIN ANGLE DEGREES		
Pounds	Kg	0°	6°	12°	0°	6°	12°
1.000	454	35.500	27.100	21.500	28.500	23.900	20.500
3.000	1.360	68.850	55.300	45.400	56.060	44.900	37.300
6.000	2.721	120.350	99.200	79.850	99.400	80.900	66.300
8.000	3.628	152.200	123.300	101.000	124.600	101.100	83.300
10.000	4.535	181.100	146.600	119.500	149.600	123.200	99.800
12.000	5.443	214.300	174.400	143.100	177.200	144.050	116.900
14.000	6.350	233.880	189.600	154.300	191.500	156.700	127.300
16.000	7.256	257.100	209.800	171.600	212.300	171.300	141.100
20.000	9.080	298.300	245.100	199.400	249.300	205.100	163.050
30.000	13.606	401.200	316.600	255.500	325.200	265.000	208.900

Fig. 17-2. Offdrill anchor characteristics.

AUXILIARY EQUIPMENT   191

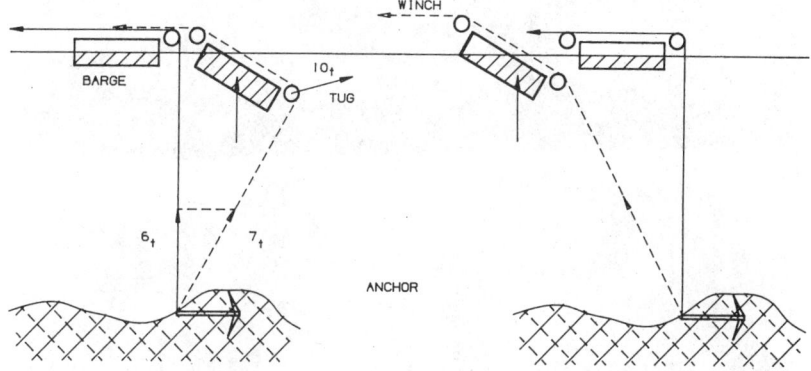

Fig. 17-3. Anchor breakout techniques. Courtesy: R. Van Den Haak.

sensus among United States operators is that the limited reach and more frequent setting as required by anchor booms tip the scales in favor of the anchor barge. The cost of anchor booms is considerable, along with the demands made on the hull for additional freeboard and stability, so that most operators are easily dissuaded from booms.

The anchor boom winch line connects to the anchor pad eye or lifting bracket (not the shank shackle which is connected to the swing winch) through a guide sheave forward of the boom foot pivot, and on through a sheave at the top of the boom, and down

Fig. 17-4. Anchor booms. Courtesy: Ammco.

Fig. 17-5. Winch, 4 drum. Courtesy: Mobile Pulley & Machine Works.

to the anchor. If an anchor has a holding capacity of 15 times its weight, and 30 percent of the holding capacity is the breakout force required, the line pull of the anchor boom winch (or anchor barge) should be 4.5 to 5 times the anchor weight. A good rule of thumb is 5.

The anchor hoisting wires must be coordinated with the swing wire speed in order to play out or in as the dredge swings. The anchor hoist line speed needs to be 10 to 15 percent faster than the swing speed with modulated clutch slippage provided for automatic speed adjustment. Obviously, port and starboard drums are required to service their respective anchors. Drives can be separate for each drum, or both drums can be added to the forward winch with clutches and brakes, making it a 5-drum unit, with a single motor.

SPUD HOIST WINCHES

The usual spud winch is a 2-drum unit, one for the port, the other for the starboard spud. If a stern mooring swivel is required, a third drum is added.

AUXILIARY EQUIPMENT

The line pull of the spud winch is generally equal to the weight of the spud, but since the hoist normally has 2 parts, this provides twice the pull to draw it from the clinging bottom. In sticky material this can be borderline, but is generally sufficient. The line speed should result in lifting the spud at a minimum of 30 feet per minute to conserve time, and some operators prefer a faster speed. Since horsepower goes up directly with speed, discretion must be used. Note that the 2-part hoist requires a line speed of 60 feet per minute to achieve the 30-foot per minute hoisting speed.

Spud winches can be replaced with hydraulic cylinders working through either wires or other arrangements such as jacking mechanisms. All spud lifts should allow for free-fall in order to penetrate the bottom for adequate holding power to resist the swing line pull.

SPUDS

Spuds are the massive steel cylinders at the stern of the dredge which vary from about 1 to 5 feet outside diameter, 1 to 3 inches wall thickness, and 20 to 100 feet in length. Their purpose is to moor the dredge while allowing it to swing and advance. Obviously only 1 spud can be down while swinging. This down spud is called the working spud, while the other is called the walking spud. The walking procedure is described in Chapter 3. The length of the spuds required is a function of the digging depth plus bottom penetration when the spud is dropped. Spud outside diameter and wall thickness are determined from the forces applied through the top and bottom spud gates (mounted on the hull) and resisted by the spud point penetrating the bottom. The stresses are created by the swing wire force, plus wind and current.

Spuds are notoriously troublesome unless conservatively designed and competently fabricated. Proper preheat of the massive joints for welding, plus elimination of notches and welding cracks are essential to good performance.

Square cross-sectional spuds have been used where rolling capabilities of heavy walled tubes were unavailable. Since the square spud cannot rotate in its gate or keeper, its point rotates in the bottom as the dredge swings. This has the disadvantage of wearing the point, and jeopardizing the stability of the spud's position in the bottom.

Hoist-operated spuds can be top lifted or collar lifted. In either case, 2-part hoists can be used by mounting a sheave in the top of the spud for top lift, or in the collar for collar lift. The collar is a loose fitting, rugged fabrication encircling the spud. When the collar is lifted, it transmits its force through a heavy pin which passes completely through the spud in holes arranged at intervals along the spud length. The advantage of the collar lift is the moderate height of the spud frame structure required, since with the collar, the spud can extend above the frame. For deep digging dredges, the frame could tower as much as 100 feet above the deck for top lift, an unwieldy and uneconomic condition. On the other hand, once the high frame cost is absorbed, the troublesome shifting of lift pins is precluded as digging depth changes. Also, spud walls are not violated by the pin holes which can result in leaks and stress failure. Top lift should be considered for moderate digging depths.

SPUD CARRIAGE

The dual or walking spud arrangement previously described is the norm in the United States. In Europe, many modern dredges have utilized the spud carriage, where the working spud is mounted on a traveling carriage, generally activated by a hydraulic cylinder. The advent of the bucket wheel in the United States has forced the use of the spud carriage, and higher dredge efficiencies have resulted. Fig. 17-6 depicts the conventional spud, and in Fig. 17-7 the spud carriage production diagrams show a significant plus for the carriage. Fig. 17-8 shows a bucket wheel dredge assembly, complete with spud carriage or traveling spud.

WIRE ROPE

Wire rope is a high wear, high cost, maintenance item on a dredge. Proper selection and application can help control costs, which can be exorbitant otherwise.

There are several grades of wire, but only *plow steel* and *improved plow steel* are acceptable for dredge application. Preformed,

AUXILIARY EQUIPMENT

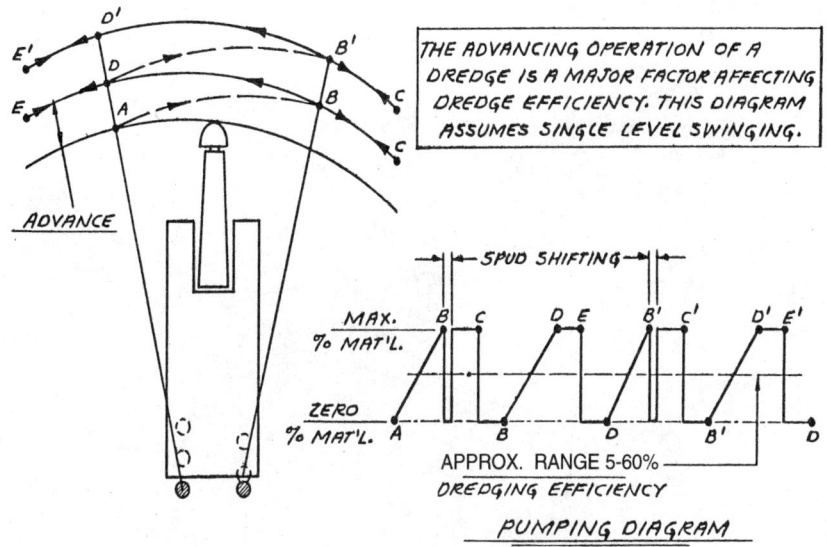

Fig. 17-6. Production diagram, walking spud.

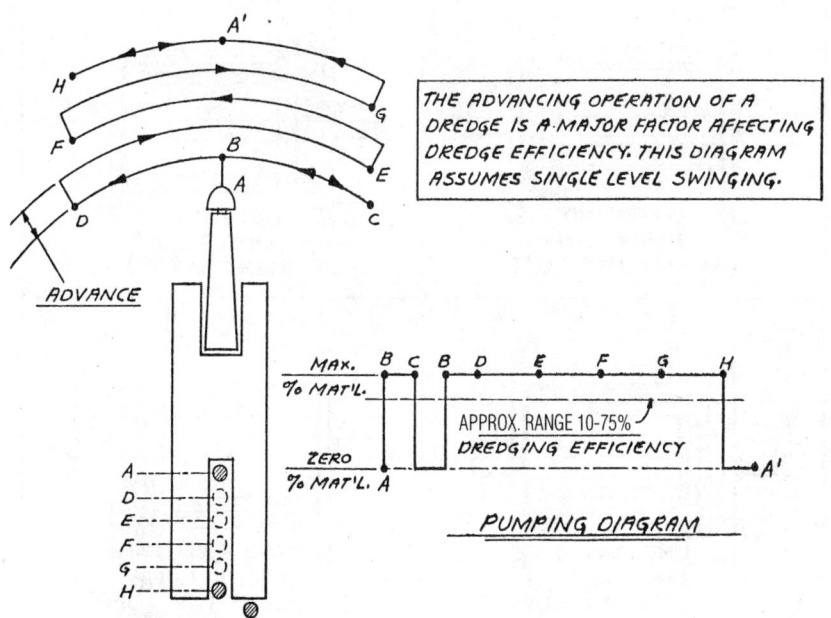

Fig. 17-7. Production diagram, spud carriage.

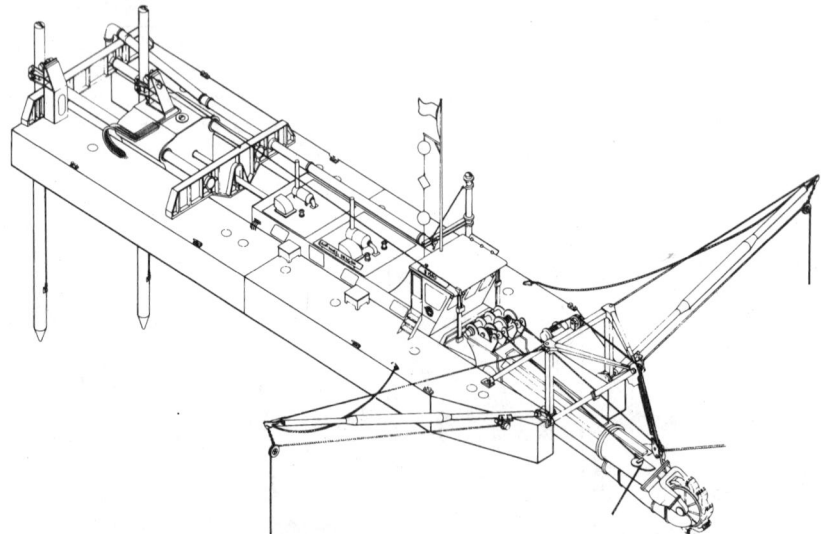

Fig. 17-8. Bucket wheel dredge with spud carriage. Courtesy: Ellicott Machine Corporation.

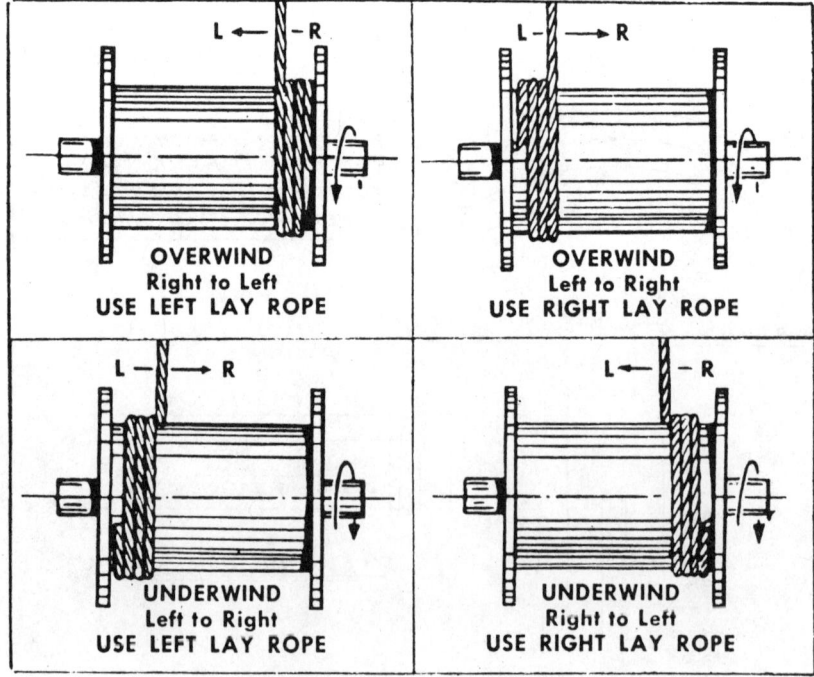

Fig. 17-9. Wire rope lays.

AUXILIARY EQUIPMENT 197

6 × 19 improved plow steel is recommended. This longer lasting wire is somewhat more expensive, but the extra cost becomes insignificant in view of the cost of downtime required to change wires.

The 6 × 19 designation refers to a wire rope consisting of 6 strands of 19 filaments each. If the filaments are laid to the left, and the strands to the right, the rope is said to be right regular lay; if reversed, the rope is left regular lay. The Wire Rope Institute recommends right and left lays be wound on winch drums as shown in Fig. 16-9. Improper winding of a lay may cause the rope to foul, crimp, and snarl, shortening its life.

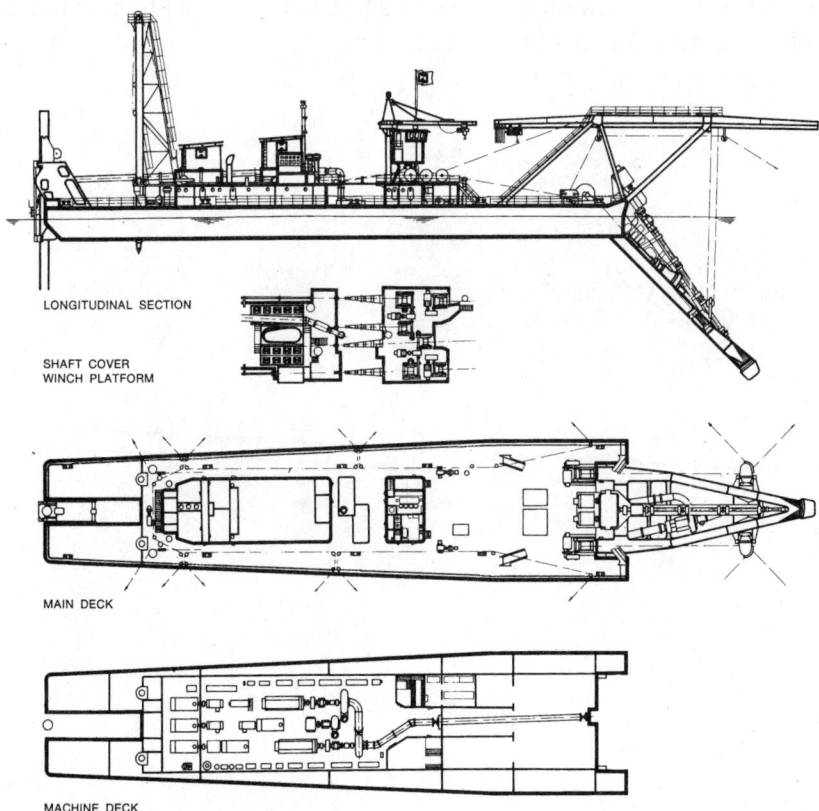

Fig. 17-10. Dredge *Cutter PH V* plan. Includes three pumps plus cutter lineshaft. Courtesy: Philipp Holzmann.

Sheaves and drums should not have diameters less than 15 times the wire diameter, and should be greater preferably. Small diameters and reverse bends are highly deleterious to wire rope and are to be avoided.

Wire rope should have a breaking strength of at least 2.3 times the maximum working force. This provides a higher safety factor for the average working force, since the maximum force is only applied intermittently.

SUMMARY

Properly designed and operated auxiliary equipment is essential to the efficient operation of a dredge. The operator will do well to reflect upon the principles of such equipment in this chapter when selecting winches, anchors, spuds, etc.

Chapter 18

INSTRUMENTATION AND AUTOMATIC CONTROL

The hydraulic dredge is a unique tool in one unfortunate respect; the operator is unable to see his production, and therefore, is unable to optimize his operations by visual observation. Excavation of unseen terrain goes on beneath the water surface, while the entrained solids pass through an opaque pipe to the deposition area, thousands of feet or perhaps miles away. Even if the operator could see the distant discharge, the time lag between the controlled suction conditions and the arrival of the slurry at the deposition area would probably cause more confusion than enlightenment. Therefore, instruments which concurrently inform the operator what is happening in the dredging process are essential to efficient operation. Also, there are instances where simple automatic controls are desirable and justifiable.

While the personal computer is not necessarily an instrument installed on the dredge, it is so important to successful dredge management that it should be a primary consideration. See Chapter 20 for further discussion.

DEFINITIONS

A *sensor* is the device that senses the parameter to be measured, e.g., the curved bourdon tube in a pressure gage, which tends to straighten with higher internal pressure.

An *indicator* is the readout that displays the value of the measured parameter. In the pressure gage example, as the bourdon tube straightens, it rotates an indicating needle on a calibrated dial.

A *controller* is a device which accepts the sensor output, compares it to a set (desired) value, and transmits an error signal to a device capable of adjusting the measured parameter.

A *positioner* accepts the controller error signal and positions a device, which results in a change in the measured parameter, e.g., the throttle of an engine driving the pump creating the velocity in the slurry line.

DURABILITY

Dredge instruments exist in a generally hostile environment. Devices that intrude into the slurry pipe are particularly vulnerable to wear and/or damage. The marine atmosphere and the vibrations of a hard-working dredge combine to demand ruggedness of instruments beyond that of most other applications. Components should be solid state where possible, shock-absorber mounted, and protected against the elements. Spare parts should be available for all instruments.

ACCURACY

Accuracy of dredge instruments is desirable but not necessarily essential. A sensor that habitually reads 10 percent low, e.g., a ve-

Fig. 18-1. Lever room control stand. Courtesy: Ellicott Machine Corp.

locity meter, can be a valuable guide to the operator as long as it reproduces its readings consistently. The extra cost of achieving high accuracy on dredge instrumentation is probably not justifiable.

DREDGE POSITION

With satellites orbiting the world, the location of the dredge on the surface of the earth within 5 meters or less can be ascertained by Global Positioning Systems (GPS). However, the volume of soil within 5 meters on either side of a channel could represent a sizable percentage of the total project cubic yards; so while the GPS can locate the dredge with reasonable accuracy, a benchmark agreed to by the owner and contractor is desirable for precision.

Radio signals from fixed transmission sites can be used to position the dredge with good accuracy. Many dredges still use line of sight targets and/or laser beams for guidance. The operator has various options today, but it should be emphasized that he should select and become proficient with one or more, since many a cubic yard has been pumped without compensation by an improperly positioned dredge.

Either a gyro or magnetic compass is required to indicate the swing of a cutterhead dredge. The gyro is expensive; less costly devices are available that accurately measure the swing angle of the dredge, and are generally sufficient for project control. An audible warning signal on the swing system is useful to the leverman (perhaps five degrees prior to the maximum angle of the swing). This prompts him to reverse the swing and avoid over swinging.

In recent years, sophisticated systems showing visual displays of the dredge and project have become available. These systems seem to be limited only by the imagination of the developer, i.e., until cost is considered. They can show an overall view of the project, completed and uncompleted dredge reaches, the location of the dredge, its position in its swing cycle, the location of the cutter, etc. It has the capability of quickly orienting management to the project status, and a readout can be located in the home office.

These systems are unquestionably appealing, but they are probably more useful to management than to the operator. Certainly,

202   DREDGING IN PRACTICE

they are costly, and the justification is dubious from the leverman's viewpoint. The author suggests that such systems may be justifiable under some circumstances, but a careful economic analysis is recommended.

SLURRY SYSTEM

Traditionally, the dredge has had vacuum and pressure gauges for the slurry system. The vacuum gauge, which measures the losses in the suction line, allows the maximization of solids while avoiding cavitation. The pressure gauge, which measures the losses in the discharge line, provides information vital to the avoidance of plugging the line. These gauges are essential instruments, and should be installed carefully to avoid fouling of their sensor lines.

If there is a ladder pump, a combination vacuum-pressure gauge is needed on the suction. This can be a gauge in the lever room which is mounted on a metered air line which connects to the suction line just ahead of the ladder pump. The air in the tube must build up a pressure equal to the hydrostatic head at its ter-

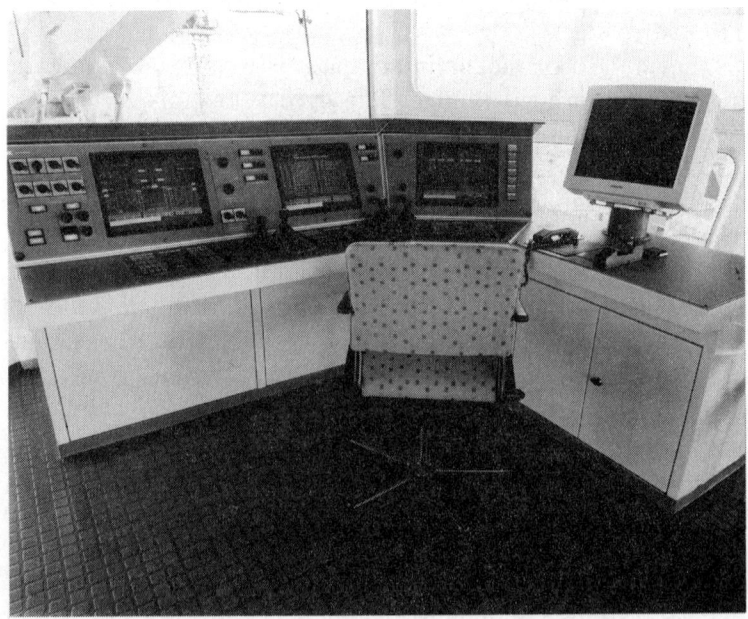

Fig. 18-2. Dredge control console. Courtesy: IHC.

INSTRUMENTATION AND AUTOMATIC CONTROL 203

mination in order for the air to escape. In the event of a vacuum, the metered air in the tube will be drawn down to indicate less than atmospheric pressure.

The first dredge pump in the hull also requires a combination vacuum-pressure gauge on its suction if used in conjunction with a ladder pump. Otherwise, a vacuum gauge only is required, but it should have a simple petcock air bleed at the control stand to allow intermittent or continuous purging of the vacuum line to avoid plugging and inconsistent readings.

The second dredge pump, if used, requires a pressure gauge on the suction, and one on the discharge to indicate total line pressure. This latter gauge would, of course, be used on the discharge of the first dredge pump if a second dredge pump were not used. This gauge is normally mounted on a stand pipe and equipped with a water purge system to avoid plugging because of ram offs in the dredging system.

Each pump requires RPM indication with the possible exception of the ladder pump which can normally be constant speed. Torque or horsepower indication is desirable for each pump if diesel driven to avoid overloading, unless the governor is the type that limits the torque. Such torque indication is highly desirable or even mandatory if electric drives are used in order to avoid overload shutdowns.

Automatic slurry velocity control is so inexpensive relative to its value to the operation, that it should be included on most hydraulic dredges. This control has contributed to surprisingly large increases in production rate, as well as savings in fuel costs (see Chapter 9 for dredge cycle explanation). The control is simple. As an example, a strap-on doppler velocity meter senses the slurry velocity, sending a corrective signal to an engine throttle controller, functioning through an air-powered positioner. This positions the throttle to adjust the pump speed up or down as required to keep the slurry velocity in the selected range for efficient transport of solids. With the all-important slurry velocity assured, the leverman is free to devote his time to other factors which affect production.

Traditionally, the leverman has controlled his slurry system by vacuum and discharge gauges, not because they were an accurate indication of dredge performance, but because they were the best indication available. Today there are several companies who offer production meters which provide instantaneous indication of specific gravity, velocity, production rate, and integrated production. Knowledgeable operators are aware that the intelligence provided by such meters has the capability of increasing production substantially. A dredge which operates today under variable conditions, without such intelligence to optimize its conditions, labors under an unnecessary handicap. The production meter is highly desirable, and considered by some to be mandatory.

The production meter combines the slurry velocity and specific gravity meters to achieve production rate. Many operators have reported difficulty maintaining the specific gravity indicator, normally a nuclear device emitting controlled radiation, directed to an ionization chamber. While desirable and helpful, the SG indicator, in the author's opinion, is not as essential as the personal computer or the slurry velocity control to the successful management of the hydraulic dredge.

In recent years, manufacturers have offered equipment purportedly capable of automating the dredge cycle. The underlying principle is to sense SG of the slurry in the suction line, and to optimize

Fig. 18-3. Production meter. Courtesy: Ellicott Machine Corp.

it by automatically speeding or slowing the dredge swing rate with a control system.

DREDGE CYCLE AUTOMATION

In recent years, several manufacturers have offered equipment purportedly capable of automating the dredge cycle. Unfortunately, these efforts at automation have not all been successful, and, indeed, it is doubtful that any models have been the unqualified success their manufacturers had hoped for them. The major contributing factors for the lack of success have been the widely varying topography and consistency of the bottom where employed; failure to understand the dredge cycle thoroughly; and the considerable complexity of the necessary circuitry which demands much technical competence and troubleshooting capability.

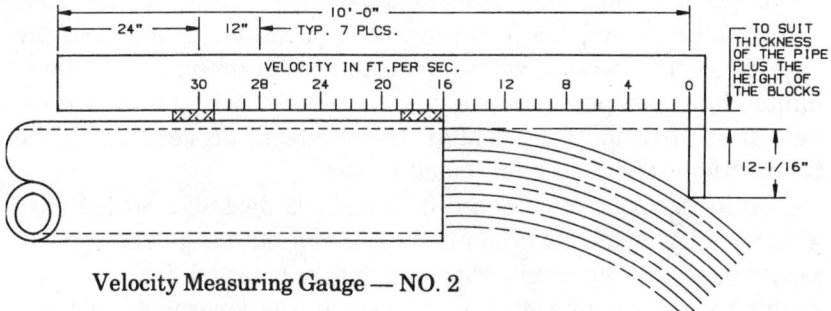

Velocity Measuring Gauge — NO. 2

A fairly accurate measure of the velocity in the discharge pipe can be made with the gauge as indicated above, providing the last section of discharge pipe is leveled to a horizontal plane. This can usually be accomplished by wedging the last section of pipe to suit. Place two blocks near the end of the discharge pipe as shown on the sketch. The top surface of the blocks must be in line with the inner surface of the discharge pipe. Make a straightedge about 10 feet long, and provide it with an offset at 90° (right angle), such that the offset is 12-1/16 inches plus the distance from the inner surface of the pipe to the bottom face of the straightedge, as shown above. Graduate the straightedge with four divisions per linear foot (each subdivision being 3 inches), as shown above, the zero being opposite the inner face of the offset. Slide the straightedge back and forth until the offset just touches the upper surface of the water discharge, and the velocity in feet per second can be read directly by taking the subdivision opposite the end of the pipe. In the sketch shown above, the velocity of discharge is 16 feet per second.

Fig. 18-4. Velocity measuring gauge.

There are few cutter dredges today that can justify full automation. Such controls are expensive to install and maintain, too much so to buy leisure time for the leverman. The only real payoff lies in increasing the productivity of the dredge, but this is largely available to the leverman with the less costly PC or the production meter intelligence with which he manually "closes the loop." It is questionable procedure to attempt the quantum jump from a "blind" dredge (i.e., no PC or production meter) to a completely automated dredge. It is more advisable to allow the operators to learn the dredging cycle thoroughly with the PC simulation. From this higher level of knowledge, the decision concerning automation can be made more intelligently.

CUTTER MODULE

Torque or load indication is mandatory on the cutter. In the case of hydraulic power, the fluid pressure indication (or amps in the case of electric power) will suffice. Almost all dredges have been limited by cutter power at one time or another. Without the torque indication to guide the operator, the dredging process would be frequently interrupted by a stalled cutter.

Cutter RPM or other speed indication is desirable where variable speed is available. There are optimum cutter speeds for varying materials, but even when digging a material where speed seems to be insignificant to performance, the leverman can slow down to the lowest practical speed to reduce wear and tear. In the absence of speed indication, an "on" light is desirable. The author, a would-be operator, can attest to the embarrassment caused by trying to swing the dredge with the cutter not running.

DIGGING DEPTH

An indication of digging depth is mandatory. Traditionally, this has been accomplished by the use of a pointer on a painted gauge located on the forward frames of the dredge. The pointer is wire connected to the ladder to provide positive indication of the depth as the ladder descends. The bubbler tube, previously described, is a satisfactory alternate.

INSTRUMENTATION AND AUTOMATIC CONTROL

The depth indicator should be calibrated for the depth of the suction intake at the back-ring of the cutter. Digging depth is not the extreme reach of the cutter since only the material in the immediate vicinity of the suction intake can be picked up and transported.

SOUNDING

Lead sounding is still occasionally used to determine before and after dredging depth. However, acoustic sounding devices, which can provide a printed readout, are becoming more prevalent, and are widely used by the Army Corps of Engineers. Some dredges have acquired plotters, into which sounding data can be fed to provide a visual picture of the channel and the work that remains to be done.

WINCHES

All winches need torque indication to prevent stalling. Overloading can be caused on the swing winch by dragging the ladder; on the ladder winch by an excessive mud load; and on the spud winch by having the spud stick in the mud.

Winch RPM or speed indication is desirable. In an open waterway, operator disorientation is common and the ability to estimate swing speed difficult. Many dredges control swing speed simply by manually positioning the throttle, which is generally satisfactory. Precise swing speed is unimportant in itself, but along with cutter position in the bank, is the primary means to control solids content of the slurry.

SUMMARY

There are numerous devices required to control the dredge process, and others which are desirable. Automatic control of the dredge cycle is rarely justified at the present state of the art, but optimizing intelligence must be provided the leverman in order to operate efficiently. Above all, the operators should be well versed in the principles of hydraulic dredging. The PC is most effective in this regard.

Chapter 19

CALCULATING AND BIDDING THE PROJECT

An executive of one of the large dredging companies was asked what his company's greatest asset was. Knowing the importance of submitting the correct bid on all projects, he responded that while his people and his equipment were very important, he felt that their record of past jobs showing dredge capability and costs was their most valuable asset.

While there may be room for disagreement with this answer, the emphasis on knowing dredge capability and relating it to the project is not misplaced; it is frequently the difference between financial success and failure. This chapter is devoted to project calculation and dredge capability, with emphasis on elements of the job which are easily overlooked in the bid calculation.

CONTRACT DOCUMENT EVALUATION

When project documents are submitted for bids, the specifications must state the amount of material to be excavated, the location from which it is to be removed, and where it is to be deposited. The nature of the material should be specified, normally as determined by geotechnical analysis of selected, representative borings. The physical parameters of the project must be specified, including digging depth, terminal elevation, discharge line length, and dimensions of the excavation. Environmental restrictions must be specified, and schedule limitations, if any, must be established.

It is the responsibility of the dredging contractor to evaluate the owner's specifications, converting them into the time required for his equipment to complete the job. This time is then converted into costs, to which the contractor adds the necessary overhead and profit to arrive at an acceptable figure for his bid.

METHOD OF CALCULATION

Determination of the time required for a dredge to perform a specified job is complex. Frequently it is derived from historical data, but such data is fraught with risks. The change of a single parameter, e.g., swing width, bank height, or digging depth, can make a significant difference in hourly production rate. There are many other factors that influence rate, but the three cited are often not recorded in the historical data; therefore, it is essential for the operator to be able to run accurate calculations to determine the effect of any project or equipment change on the production rate of the dredge.

Hand-held calculators have often been used to perform dredge production calculations. This manual technique is time consuming, complex, and fraught with error. When the number of possible variables in a dredging project are considered, e.g., 50 concentrations of slurry, 50 digging depths, 50 terminal elevations, 15 suction line sizes, 15 discharge line sizes, 10 different soils, 40 work-face heights, etc., it is easy to see that there are hundreds of millions of possible combinations. This is the type of calculation for which the personal computer is designed.

The PC requires accurate software to perform properly. It is vital to the dredging manager that he have an electronic model of his dredge on hand when making profit or loss decisions regarding operations. If any parameter changes, he needs to be able to insert the correct value in his computer (e.g., a terminal elevation increase from 10 to 50 feet) and immediately detect the effect on production rate.

Experienced operators with in-house programming capability can develop their own software, but it is a major undertaking. PC programs are available for purchase in the dredging industry, and it is recommended that the operator procure the best and most accurate program available, cost notwithstanding. Chances are that the PC software will return its total cost many times over on the first one or two projects. The financial return on the PC and software will probably be much greater than for the dredge itself.

There are many factors to be evaluated in planning and estimating a project. The following discussion of these factors assumes the availability of a PC with competent software, able to receive the necessary data and calculate the hourly production rate essential to the project planning and execution.

EVALUATING A PROJECT

There are 17 factors which need to be evaluated in setting up a project. These factors are discussed under the headings as indicated below.

1. Material to Be Pumped

From the geotechnical data, the median grain size of the material must be determined. It may be necessary to break the project into several reaches to accommodate different soil characteristics and d_{50}, calculating separate production rates for each. A good PC program is very helpful in this regard, as it can produce data for various d_{50}s simultaneously. Figs. 19-1 and 19-2 are composite production charts for various materials from the author's PC program for a 27-inch dredge. Fig. 19-1 is a "D" dredge (no ladder pump) and Fig. 19-2 is an "L" dredge (with ladder pump). If the project d_{50}s fall between the materials plotted on the charts, interpolation can provide reasonable results.

Free-flowing material, which flows readily to the suction intake, provides an improved dredge efficiency over standing material, as shown in Fig. 3-2.

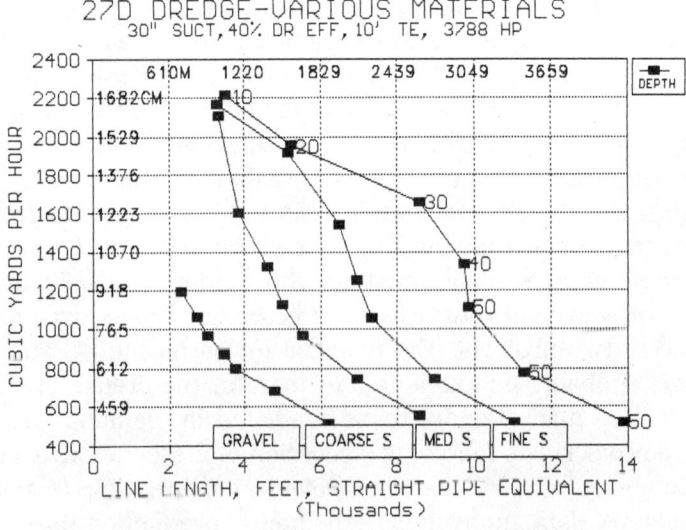

Fig. 19-1. Production chart: 27-inch dredge with 30-inch suction.

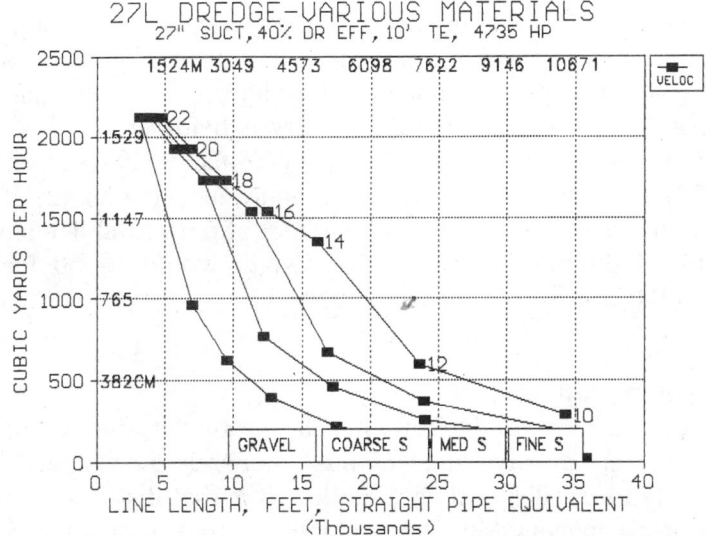

Fig. 19-2. Production chart: 27-inch ladder pump dredge.

2. Digging Depth

Digging depth has been established in Chapter 6 as one of the significant factors affecting capacity of the dredge. The maximum depth requirement of the project must be ascertained in order to determine if the dredge has the necessary ladder length, and whether or not a ladder pump is necessary to make the project economically viable. A project of significant magnitude would probably justify a ladder pump when average digging depths are greater than 20–30 feet. The production chart of the "L" dredge should be checked against that of the "D" dredge for production rate differential.

Average digging depth on each type of material to be dredged is a significant parameter for the estimator. A reasonable rate estimate can be obtained from the production charts for the average depth. Tides must be considered, and a good tidal datum is important to the leverman to avoid over- or under-digging.

3. Terminal Elevation

Normally the pipeline goes ashore and discharges about 10 feet above water level. Therefore, it has been standard practice to build into the production charts a 10-foot terminal elevation. Occasionally, however, the terminal elevation requirement is higher, and when this occurs, the frictional equivalent in line length must be added to the pumping distance for the chart to be applicable. For example, if a 1.25 specific gravity slurry is being elevated 60 feet at its termination, then 50 feet elevation (60 feet − 10 feet terminal elevation in chart) must be added in equivalent line length. If the friction loss is 5 feet of water per 100 feet of horizontal line length, then the additional equivalent line length would be 50 feet × 1.25 × (100 feet/5 feet) = 1,250 feet.

4. Discharge Line Length

If order to use the production chart accurately, it must be understood that the discharge line length is not merely the distance from the dredge to the disposal area. Rather, it is that distance plus the length of the maneuvering loop behind the dredge, plus the equivalent line length of all fittings such as flap valves, ball joints, swivel elbows, tapered joints, undersized pipe, etc. The pump "feels" only the resistance of the system, and its output is the same whether the resistance is created by 100 feet of 20-inch pipe or 240 feet of 24-inch pipe or ten 24-inch long radius 90° elbows. But, in order to use the convenient production chart, all resistances, including terminal elevation, must be converted into *equivalent length* of the dredge discharge pipe size.

Any good hydraulic reference book can provide equivalent line lengths (or losses expressed in terms of velocity head) for fittings. Cameron Hydraulic Data[4] is good, and there are others. In the absence of better information, a reasonable estimate would be the addition of 10 to 20 percent of the actual distance from the cut to the disposal area to obtain equivalent line length for pricing purposes.

The maximum project line length is significant to determine whether it is feasible to operate without a booster pump. However, an even more significant figure is the *average* line length, which provides a reasonable basis for calculation of the entire project.

5. Cutter Capability

The cutter functions as an excavator, feeding solids to the suction mouth or intake. If the cutter has less capacity than the hydraulic

CALCULATING AND BIDDING THE PROJECT

transport (slurry) system, the dredge rate must be down-graded accordingly. See Chapter 13 and Fig. 13-10.

6. *Height of Work Face* (Bank Height)
The height of the work face is the height of the earthen bank the cutter is excavating, notwithstanding its position with respect to the water level. The distance from the bottom cut to the top of the bank multiplied by swing width and cutter length determines the cubic yards to be excavated by merely lowering the cutter without advancing the dredge. This obviously affects dredge efficiency.

Generally, the low work face is encountered on maintenance dredging, where the material is a free-flowing sand or silty material. Here an effective technique to improve dredge efficiency may be to over-advance and over-dig. This results in bypassing some material, and digging other material below specified grade. However, it has the effect of increasing the work face or bank height from perhaps 50 percent of the cutter diameter to 100 percent, increasing dredge efficiency by 10 percent and production by 20 percent. Since the bottom material is free-flowing, it has a tendency to level out the furrows created by the technique, leaving the bottom in a condition as though shorter, shallower, less efficient advances had been made. When "chasing the material" in a low work face, the efficiency of the dredge is greatly improved by a traveling spud carriage. Bank height has a significant effect on dredge efficiency. See Fig. 3-2.

7. *Swing Width*
Dredge operators have noted their production rate increases with the width of the channel being cut in a single swing. The wider channel allows more material to be dredged before the lost time of the dredge advance occurs; therefore, DE rises, increasing production rate. Some operators have added a spud barge to the stern to lengthen their dredge, allowing an increased swing width for a given swing angle. This is an effective measure where the channel design allows it. See Chapter 3.

8. *Type of Advancing Mechanism*
The walking spud is the most common advance mechanism for cutterhead dredges. It is slower than the spud carriage mechanism

or the Christmas Tree arrangement (mooring cables). The last two mechanisms are considered as having equal effects on DE, while the walking spud reduces DE with respect to the other two.

9. Dredge Efficiency

DE is largely a function of soil type, bank height, swing width, and the dredge advance mechanism. Operator skill can also be a factor, but no attempt is made to evaluate it in this book. It is feasible for the operator to lower his estimated production rate by a small percentage to account for inexperienced levermen. Reference to Chapter 3 and Fig. 3-2 will refresh one's memory regarding the method of calculating DE. Calculation by computer is highly recommended.

10. Suction Line Size

The operator normally evaluates a project against his dredge's existing configuration, e.g., suction and discharge line sizes, pump HP, and gearbox ratio. While it may be possible to improve performance of a dredge on given project conditions by changing any of the four items above, the most likely improvement (and generally the lowest cost) involves changing the suction line size. This was discussed in Chapters 7 and 8, Figs. 7-1 and 8-2. In brief, if the suction line is larger than the discharge, the dredge will produce well on short lines; if the lines are the same size, the dredge will be relatively more productive on long lines. It is not good practice to have the suction line smaller than the discharge for fear of plugging the discharge line (see Chapter 4). With the PC, it is a quick exercise to compare the effects of different suction line sizes. See Fig. 19-3, which shows the effects of 20-inch and 18-inch suction lines on a dredge that is otherwise identical.

11. Hourly Production Rate

The production rate of the dredge can now be determined by the PC using the proper inputs for material, line sizes, digging depth, terminal elevation, dredge efficiency, pump HP, and impeller tip speed. Maximum discharge line length is calculated automatically. Cutter capability becomes a factor only if it is less than that of the slurry system.

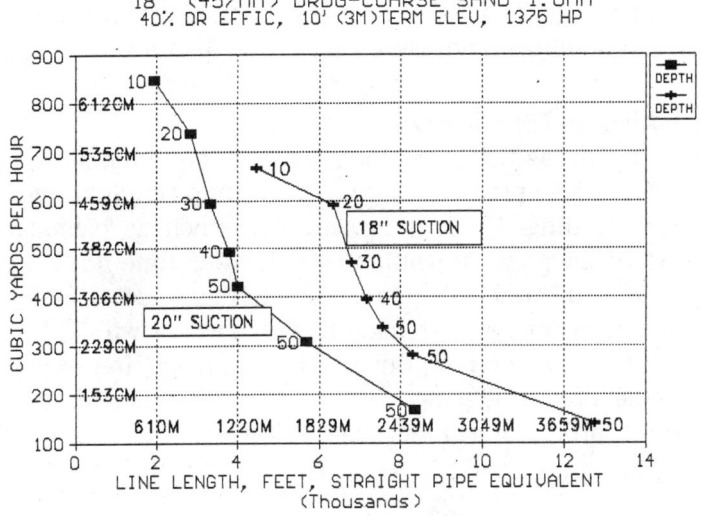

Fig. 19-3. Production chart: 18-inch dredge, 20-inch vs. 18-inch suction sizes.

12. Total Project Yards

The total cubic yards in the project must be ascertained from project data, generally from contour and prism information supplied by the owner. Consideration of soil "swell or shrink" factors, payment on cut or fill measurement, plus intrusion of new material after dredging should be made. Also, different methods of sounding should be analyzed since they can provide widely divergent results, especially in soft materials.

The total cubic yards calculated from the contract documents seldom, if ever, represents the actual yards pumped by the dredge. To avoid expensive re-dredging of an area (because of failure to clean up the bottom or sides of the design prism adequately), the leverman learns quickly that over-dredging is the economical procedure. The cubic yards involved in this over-dredging can be considerable. The author has seen over-dredging as high as 50 percent, but a more reasonable average would be 15 percent. Some contracts allow payment for over-dredging within certain limits, but others allow none at all. It is imperative that the bidder evaluate the necessary over-dredging properly in order to arrive at the time and cost required to perform the project. He should calculate the "design" yardage from the contract documents, add the overdredging percentage, and use the result as his total project yards.

13. Production Time Required

Production time is that time spent actually pumping slurry. Some operators confuse production time with pump running time which can be misleading. Production time is defined as calendar time minus downtime, and downtime is defined as time not scheduled plus unplanned, nonproductive time such as unscheduled repair, spoil area problems, etc. Frequently, the operator will let his pump continue to run to avoid repriming while making minor repairs; but without the dredge digging, there is no production. If this were considered production time, the following equation would be inaccurate.

$$\text{Production time (hours)} = \frac{\text{total project cy}}{\text{production rate cy/hr}} \qquad [Equation\ 19\text{-}1]$$

Since we have previously established the total project cubic yards and the dredge hourly production rate, production time in hours becomes a simple calculation.

14. Calendar Time Required

Unfortunately, from the operator's viewpoint, his costs continue 24 hours a day, 7 days per week on a full calendar basis, whereas his hourly production rate is effective only while digging. Some costs are reduced when the dredge is not running, e.g., fuel and wear, but many continue such as rent, salaries, etc. Therefore, it is essential that a realistic relationship be established between calendar time and production time. The contract and liquidated damages are based upon calendar time, so that in the final analysis, we must return to calendar time.

$$\frac{\text{Production time}}{\text{Calendar Time}} = \text{percent production time} \qquad [Equation\ 19\text{-}2]$$

$$\text{Or Calendar time} = \frac{\text{production time}}{\text{percent production time}} \qquad [Equation\ 19\text{-}3]$$

Or Production time = calendar time
× percent production time $[Equation\ 19\text{-}4]$

CALCULATING AND BIDDING THE PROJECT 217

Perhaps the most widely accepted estimate of percent production time is 75 percent of calendar time. Many operators claim they operate 22 or 23 hours out of 24, but in the writer's experience, there is as much exaggeration regarding dredge operating time as there is about gas mileage. A realistic figure based on industry experience provides 18 hours/day of production time, or 75 percent. A taut, efficient operation may become proficient enough to achieve 83.3 percent production time based upon the following scenario:

$$
\begin{aligned}
7 \text{ days/wk} \times 24 \text{ hrs/day} &= 168 \text{ hrs/wk} \\
1 \text{ 8-hr shift planned maintenance} &= -8 \text{ hrs/wk} \\
1 \text{ hr/shift unplanned downtime} &= -20 \text{ hrs/wk} \\
\text{Net production time} &= 140 \text{ hrs/wk}
\end{aligned}
$$

$$\frac{\text{Production time}}{\text{calendar time}} = \frac{140}{168}$$

$$= .833 \text{ or } 83.3 \text{ percent production time}$$

Few operators achieve this proficiency on a consistent basis. Many run 50–70 percent, and some even less. Over-estimation of percent production time is a major cause of project failure.

15. Trash vs. Production Time
Trash, defined as any oversized material adversely affecting the hydraulic transport system, can be so troublesome as to be worthy of special mention. The author has seen trashy conditions so bad as to cost the dredge 50 percent of its production or even to shut down the operation until special provisions could be made in the cutter, suction line, and pump. Water hyacinths, cattails, roots, stumps, etc., can be so deleterious to a dredge's performance (particularly for small dredges up to 20″ in size) that special consideration should be given to reducing the estimated operating time by several percentage points over that mentioned above. Experience is the only guideline here. The disc type cutter has shown well in trashy situations (Chapter 13).

16. Operational Costs
The operator would be well-advised to analyze his costs to arrive at two hourly figures for his operation: (a) total costs while pump-

ing slurry; and (b) costs while lying idle. These could be applied to (a) production and (b) downtime hours, and the cost of the project calculated. It is imperative that the operator omit no costs such as overhead, depreciation, interest, wear, etc., because eventually the neglected cost will require an accounting. It is generally good practice to keep the hourly costs up-to-date, and to estimate mobilization, and other costs peculiar to the project, as additions to the project cost.

17. Bid Price

Most operators desire to bid what the market will bear, regardless of cost, but the genius of the free enterprise system is such as to maintain some relationship between cost and price. It is unfortunate that often the less knowledgeable operator who has omitted some valid costs is the successful bidder. These operators either correct their mistakes in time or experience a short life span in the industry. It is the author's considered opinion that the correct approach to bidding a job is to accurately calculate the costs, and to add an acceptable overhead and profit to arrive at the bid price.

Fig. 19-4. 16,000 HP Korean dredge, *DWPD-5*. Courtesy: IHC.

Some jobs will be lost to errant bids, or to hungry operators, but the good, consistent operator will succeed in achieving his share over a period of time.

SUMMARY

Essential to proper bidding and project calculation are the knowledge of the dredge's production capabilities under varying conditions, a realistic appraisal of production time, and an accurate evaluation of costs. This information is basic to the proper management of a dredging operation; without it, the chances of economic success are minimal. A computer with good dredge calculation software, operated by competent, experienced dredgemen, is the most effective route to efficient project planning, bidding, and operation.

Chapter 20

THE PERSONAL COMPUTER IN DREDGE MANAGEMENT

Historians will likely refer to the last two decades of the twentieth century as the era of the personal computer revolution. The analytical ability of the PC has provided management control and efficiencies never before realized.

NEED

The potential of the PC is just beginning to be realized in the dredging industry. Some companies are using it successfully, but far too many are still struggling with old methods that penalize their profits and jeopardize their future. The purpose of this chapter is to describe the merits of the PC to all operators and/or users of hydraulic dredges.

There is a large number of variables that affect the performance of a dredge. Among these are the nature and particle size of the soil, size of the suction and discharge lines; digging depth, pump HP, impeller diameter and speed, line length and terminal elevation; work face height, width of dredge swing, and whether or not the dredge is equipped with a ladder pump. If one recognizes that each of these variables must function with 50 different concentrations of slurries between 1.0 SG and the maximum practical slurry of 1.5 SG (each of which has different rheological or flow characteristics), it becomes obvious that there are millions of combinations to calculate. It is extremely difficult, time-consuming, and error-prone to keep all these variables in proper order for a manual calculation; however, for a personal computer, it is easy, quick, and accurate.

SOFTWARE

Of course, the computer must have a competent software program to direct it. If the software program is accurate, the computer becomes an electronic analog (model) of the dredge. The program must allow for the entry of all physical characteristics of the dredge, the soil, and the project. While this sounds complex, it is actually quite simple. With good software prompting for the necessary entries, there is assurance that nothing is overlooked (the most common fault). It is significant that nothing is required as an entry to the PC program that is not required for an accurate manual calculation of the dredge performance, and the PC result is quicker and more accurate. Some operators have been fearful of their perceived complexity of the computer program for dredge calculation. The truth is, the computer greatly simplifies dredge calculation, saving much time in the process.

The first requirement of a good software program is a proven technical data base, i.e., a table of values showing required slurry transport velocities, friction coefficients, and pump characteristics. For example, Fig. 4-1 shows slurry velocities required to transport various solids at each concentration (SG). The Che chart, Fig. 1-7, shows the coefficient necessary to adjust for loss in pump head and efficiency as caused by the amount and nature of the solids in the slurry. Fig. 10-2 shows the Hazen-Williams friction coefficient for the specific gravities of various soil slurries. Also, the table must include the operating characteristics (head, flow, and efficiency) for the dredge pump as it performs on *water*. The computer then converts the water data to that of the appropriate *slurry*.

All of these data must be included in the software by a table that exceeds 200,000 bytes. Such an extensive table emphasizes the complexity of manual calculations; however, by virtue of automatic "look-up" formulas in each cell that require a table value, the computer finds the values quickly and unerringly. The complex procedure is made to order for the PC.

ACCURACY

How accurate is the calculation of dredge capacity by computer? The answer is, "as accurate as the input data and the program simulation allow." A proven program can provide very good results, well within an acceptable 10 percent range. The author has seen instances with less than a 3 percent deviation from actual results; however, even without such accuracy, the program has tremendous value for the operator. If the user entered incorrect data and the program were off by 20–25 percent, it would still show the relationship between conditions the operator needs to compare, e.g., the percentage reduction in capacity occurring when the line length is increased by 5,000 feet; when digging depth is doubled; or when coarse sand rather than fine is encountered. Under these conditions, decisions regarding the addition of a ladder pump or booster pump could be just as valid as though the predicted capacity were 100 percent accurate. Ultimately, perhaps the most telling answer to the accuracy question is this:"The computer is more accurate than manual calculation, and quicker and cheaper, too."

"D" VS. "L" DREDGES

The goal of the software is to be able to calculate the hourly capacity of any hydraulic dredge on any material, under any project conditions. It quickly becomes obvious to the knowledgeable dredgeman/programmer that two distinct programs are required, one for the dredge with a ladder pump (type "L" for ladder), and one for the dredge without (type "D" for dredge pump only). The barometric limitation of the dredge without the ladder pump complicates its calculation (and operation) in a major way. The computer shows clearly the simplicity and advantages of the ladder pump dredge over one not so equipped.

PC PROGRAM OUTPUT

Figs. 20-1 through 20-6 show typical operating data available from a PC program for dredge calculations. Fig. 20-1 is the spreadsheet for an 18-inch "D" dredge, providing optimizing data as an operating guide. Figs. 20-2 and 20-3 are two of the several production charts derivable from the spreadsheet data. Fig. 20-4 is the spreadsheet for an 18-inch "L" dredge, from which the production charts of Figs. 20-5 and 20-6 are derived.

THE PERSONAL COMPUTER IN DREDGE MANAGEMENT

DSUC	DPTH	SGMA	D.EF	SGAV	VSUC	GPM	HSL	EFS	VDIS	C	F100	DD	TE	HSUC	CY/HR	LL	HP
FINE SAND .01 MM																	
20	10	1.50	0.40	1.200	17.09	16736	196	0.77	21.1	119	7.73	18	10	19.4	904	1934	1080
20	20	1.50	0.40	1.200	15.12	14805	227	0.75	18.7	119	6.16	18	10	17.7	800	3001	1134
20	30	1.50	0.40	1.200	12.85	12581	276	0.70	15.9	119	4.56	18	10	16.0	679	5246	1260
20	40	1.47	0.40	1.188	10.94	10714	302	0.64	13.5	121	3.28	18	10	14.7	544	8199	1273
20	50	1.39	0.40	1.156	10.87	10648	295	0.64	13.4	124	3.09	18	10	14.6	449	8509	1240
20	50	1.30	0.40	1.120	10.59	10370	287	0.64	13.1	127	2.81	18	10	12.3	336	9188	1170
20	50	1.20	0.40	1.080	9.60	9400	277	0.61	11.9	131	2.20	18	10	8.9	203	11476	1084
MEDIUM SAND .316 MM																	
20	10	1.50	0.40	1.200	16.67	16328	196	0.74	20.6	115	7.80	18	10	18.9	882	1919	1088
20	20	1.50	0.40	1.200	14.75	14444	228	0.72	18.2	115	6.22	18	10	17.3	780	2997	1155
20	30	1.50	0.40	1.198	12.64	12380	268	0.68	15.6	116	4.59	18	10	15.7	662	5031	1237
20	40	1.40	0.40	1.160	12.60	12333	278	0.69	15.5	120	4.30	18	10	15.7	533	5621	1265
20	50	1.34	0.40	1.134	12.62	12362	280	0.68	15.6	123	4.09	18	10	15.7	447	5976	1278
20	50	1.24	0.40	1.096	11.84	11594	278	0.67	14.6	128	3.38	18	10	12.4	301	7327	1213
20	50	1.17	0.40	1.068	10.55	10331	273	0.63	13.0	132	2.58	18	10	9.3	190	9601	1139
COARSE SAND 1.0 MM																	
20	10	1.50	0.40	1.200	16.04	15701	194	0.71	19.8	112	7.65	18	10	18.0	848	1947	1081
20	20	1.48	0.40	1.192	14.50	14201	218	0.69	17.9	113	6.22	18	10	16.8	736	2842	1128
20	30	1.39	0.40	1.155	14.45	14145	230	0.70	17.8	118	5.71	18	10	16.9	593	3320	1171
20	40	1.33	0.40	1.130	14.36	14064	245	0.70	17.7	120	5.44	18	10	17.0	494	3781	1238
20	50	1.28	0.40	1.112	14.25	13958	245	0.71	17.6	123	5.17	18	10	17.0	422	4003	1225
20	50	1.21	0.40	1.084	13.00	12730	271	0.69	16.0	127	4.09	18	10	13.3	289	5841	1269
20	50	1.15	0.40	1.060	11.50	11261	267	0.64	14.2	130	3.11	18	10	10.0	182	7739	1182
GRAVEL 10 MM																	
20	10	1.32	0.40	1.126	17.76	17392	176	0.71	21.9	117	8.52	18	10	19.4	592	1510	1091
20	20	1.28	0.40	1.110	17.45	17091	188	0.72	21.5	120	7.85	18	10	19.3	508	1808	1124
20	30	1.25	0.40	1.100	17.04	16687	196	0.72	21.0	122	7.28	18	10	19.0	449	2085	1148
20	40	1.23	0.40	1.091	16.60	16254	199	0.71	20.5	122	6.94	18	10	18.8	400	2240	1148
20	50	1.21	0.40	1.084	16.24	15899	208	0.72	20.0	124	6.46	18	10	18.5	359	2570	1166
20	50	1.17	0.40	1.068	15.00	14688	237	0.71	18.5	128	5.26	18	10	15.3	270	3808	1234
20	50	1.13	0.40	1.052	13.60	13317	260	0.69	16.8	130	4.27	18	10	12.2	187	5373	1279

Fig. 20-1. Optimized spreadsheet for 18-inch dredge with 20-inch suction.

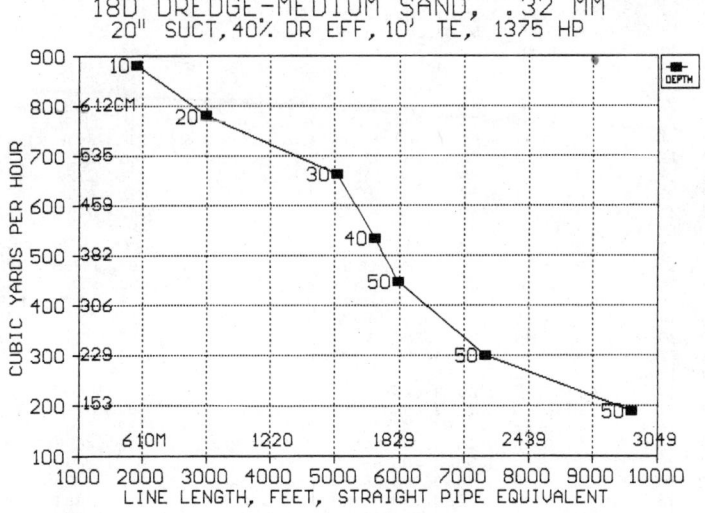

Fig. 20-2. Production chart for 20 × 18 dredge on medium sand.

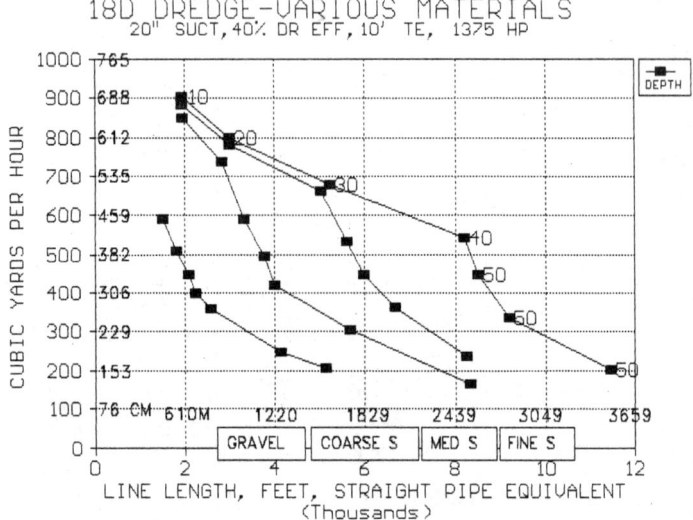

Fig. 20-3. Production chart for 20 × 18 dredge, various materials.

```
DIA VEL DPTH TEL  DEF  SGMAX SGAV  GPM    HSL  F100  EFS   CY/HR  LL     HP
FINE SAND .01 MM
18  10  50   10   0.40 1.28  1.112 7932   356  1.68  0.56  240    19554  1274
18  12  50   10   0.40 1.50  1.200 9518   380  2.72  0.61  514    12584  1499
18  14  50   10   0.40 1.50  1.200 11104  379  3.62  0.67  600    9268   1583
18  16  50   10   0.40 1.50  1.200 12690  336  4.63  0.71  685    6183   1525
18  18  50   10   0.40 1.50  1.200 14277  294  5.76  0.74  771    4111   1427
18  20  50   10   0.40 1.50  1.200 15863  260  7.00  0.76  857    2788   1366
18  22  50   10   0.40 1.50  1.200 17449  228  8.35  0.77  942    1848   1298
MEDIUM SAND .316 MM
18  10  50   10   0.40 1.16  1.064 7932   340  1.59  0.56  137    19985  1219
18  12  50   10   0.40 1.50  1.200 9518   369  2.87  0.59  514    11496  1499
18  14  50   10   0.40 1.50  1.200 11104  368  3.82  0.65  600    8456   1583
18  16  50   10   0.40 1.50  1.200 12690  327  4.89  0.69  685    5627   1525
18  18  50   10   0.40 1.50  1.200 14277  285  6.09  0.72  771    3726   1427
18  20  50   10   0.40 1.50  1.200 15863  252  7.40  0.74  857    2514   1366
18  22  50   10   0.40 1.50  1.200 17449  221  8.82  0.75  942    1652   1298
COARSE SAND 1.0 MM
18  10  50   10   0.40 1.11  1.044 7932   332  1.55  0.56  94     19952  1196
18  12  50   10   0.40 1.19  1.076 9518   338  2.33  0.59  195    13259  1367
18  14  50   10   0.40 1.50  1.200 11104  354  4.03  0.63  600    7655   1583
18  16  50   10   0.40 1.50  1.200 12690  314  5.16  0.66  685    5078   1525
18  18  50   10   0.40 1.50  1.200 14277  274  6.42  0.69  771    3346   1427
18  20  50   10   0.40 1.50  1.200 15863  243  7.80  0.71  857    2241   1366
18  22  50   10   0.40 1.50  1.200 17449  213  9.30  0.72  942    1456   1298
GRAVEL 10. MM
18  10  50   10   0.40 1.05  1.020 7932   323  1.50  0.55  43     20092  1169
18  12  50   10   0.40 1.10  1.040 9518   324  2.23  0.59  103    13381  1321
18  14  50   10   0.40 1.16  1.064 11104  326  3.14  0.63  192    9361   1463
18  16  50   10   0.40 1.23  1.092 12690  327  4.39  0.66  315    6505   1593
18  18  50   10   0.40 1.50  1.200 14277  256  7.33  0.65  771    2628   1427
18  20  50   10   0.40 1.50  1.200 15863  227  8.91  0.67  857    1729   1366
18  22  50   10   0.40 1.50  1.200 17449  198  10.63 0.67  942    1091   1298
```

Fig. 20-4. Optimized spreadsheet for 18-inch dredge with ladder pump.

THE PERSONAL COMPUTER IN DREDGE MANAGEMENT 225

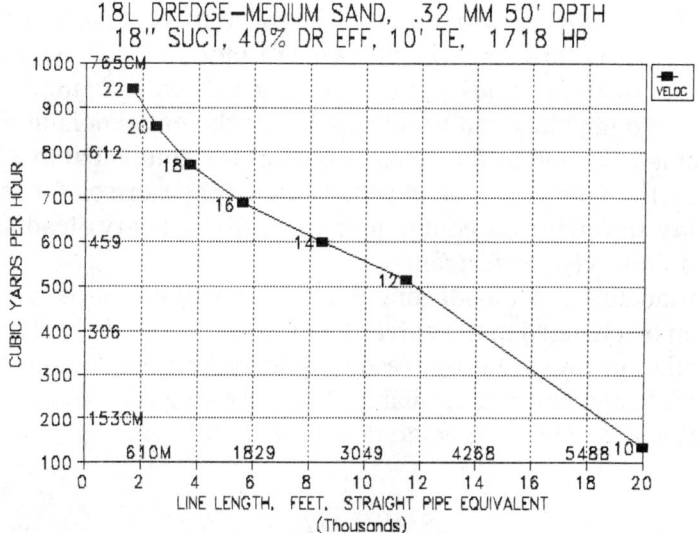

Fig. 20-5. Production chart for 18-inch ladder pump dredge, various materials.

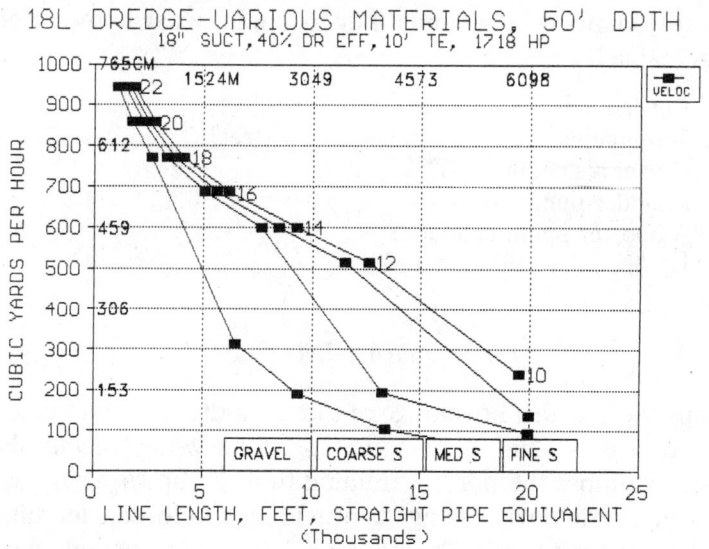

Fig. 20-6.

There are several useful ancillary procedures available for both the "L" and "D" dredge programs. One or more booster pumps can be added to the spreadsheet for all materials with a simple command. All pump data can be adjusted from the initial default values by simple insertion of the actual values of the equipment available. Further, the pumps can be checked against their drives for incompatibility (a much too common error in the industry) leading the way to improved performance.

Parameters such as digging depth, terminal elevation, and line size can be changed in the spreadsheet, with automatic background recalculation providing the results quickly. Further, a convenient "reverse" calculation is available that allows the determination of conditions to achieve specific desired result.

SIMULATION

It is difficult to overemphasize the versatility and flexibility of a good PC program. It calculates results for all entered data, but it allows the experienced operator to make adjustments at any point in the calculations to meet the dictates of his own experience. The advantages of having an electronic simulation or model, which reacts as the dredge itself, are many. The answers to "what if" questions are obtained in seconds. For example, what is the effect on dredge capacity if:

1. Digging depth is increased by 10 feet?
2. Terminal elevation is increased to 50 feet?
3. Gravel is encountered?
4. A ladder pump is used?
5. A booster pump is added?
6. Dredge pump HP is increased?

TRAINING

Perhaps the greatest advantage of the computer program is its ability to train personnel. A computer program which reacts like the dredge is almost the perfect training tool. It shows what happens when any variable is changed, providing reinforced learning experience for the trainee. The personnel training provided by a good PC program can exceed that of many years of general dredging experience. Day-to-day observation of a dredging operation is unquestionably valuable, but it does not show the mathematical relationship of each of the factors determining the capacity of a dredge, as the PC program does. Neither does it give the macro or overall view that the computer can. A company new to the dredging industry would find a quicker pay-out on its investment for good PC software than for the dredge itself.

Industrial studies have shown that successful companies can have many different characteristics, but one they have in common is the willingness to invest in the tools and training required to improve the performance of their personnel. The May 22, 1995, issue of *U.S. News & World Report* ran an article on Motorola, a leading U.S. electronics company. The article was entitled "A School For Success," and states as follows: "Training isn't the only reason for Motorola's bottom-line success, but experts contend that the company's emphasis on continuous education is a crucial advantage in today's marketplace. Training is the strongest variable we see contributing to higher returns, and its importance grows over time. And there is growing financial proof at the company that continuous learning may be one of the smartest investments employers and employees ever make."

Dredging companies with inadequate training programs would do well to emulate Motorola, utilizing the PC dredge computer program as a ready-made training tool. It is readily available for all pipeline dredges, provides continuous education, and, in the author's opinion, is the most efficient and cost-effective way to train dredge personnel.

SUMMARY: COMPUTER ADVANTAGES

The cost of the personal computer with dredge software becomes minor when considered in conjunction with its advantages. The program provides the potential of:

1. Improved project plans and estimates
2. Better evaluations of bid proposals and quoted equipment
3. Improved contract administration through the understanding of equipment capabilities and analysis of schedules
4. Improved evaluation of claims and litigation
5. Efficient, low-cost training of personnel

Any organization with a financial interest in dredges will profit by the utilization of good dredging software on the personal computer. With the PC, quicker and more accurate answers become available, and the financial statement will inevitably reflect the improvement. To waive the advantages of the PC to one's competitors in the hydraulic dredging business is economic folly.

Chapter 21

OPERATION AND TROUBLESHOOTING

OPERATIONAL ERRORS AND HOW TO AVOID THEM

Among the most expensive mistakes a leverman can make are to swing with both spuds down; to overtake and cut the lead swing wire; to allow a high bank to cave in on the ladder; to bend a spud by overswinging; and/or to get off course, or off depth.

Swinging with both spuds down results in the bending of one or both spuds, or the failure of one or both spud gates. Good practice calls for the spud gate to be designed for failure prior to the bending or breaking of the very expensive spuds. Unfortunately, it is not normally feasible to design the spuds and gates to resist full swing force, since the moment arm of the swing wire is very high as compared to the distance between the spuds. Also, it has not proved practical to provide limit switches and an interlock on the swing motor to avoid swinging when both spuds are down. While this can be done, the variable digging depths and the need to operate the winches for reasons other than swinging have precluded it. The consensus solution to the problem is proper training and rigorous procedures.

Overtaking the swing wire occurs on the back swing when the cutter is "overcutting" and "runs away." Reasonable back pressure on the trailing wire helps, but the leverman should be well aware of the hazard when he is digging hard material. He may have to decrease his advance, decrease the depth he lowers, or in very difficult material, it may be advisable to raise his ladder above the bottom on the back swing, and excavate only on the undercutting swing.

A major bank cave-in can jeopardize the dredge, not only by bending a spud or ladder, but, as cases on record have shown, by carrying it underwater due to the increased weight of earth on the ladder. The leverman should use great discretion when dredging a high bank, either above or below water. If the bank is underwater, he can terrace, rather than dig at the bottom, to allow the material to flow to the cutter. If the bank is above water, it can be seen, but the dredge reaction time may be too slow to avoid a cave-in. Here, dozers, or water monitors can be used to knock down the bank that has a tendency to cave.

When the leverman allows his dredge to get off course, or off depth, he pumps material for which he receives no pay, and he leaves material behind, for which he must make an expensive return trip. The following discussion will assume the operator has obtained the necessary surveying skills to lay out the job with required reference lines as guides, and will address the problems of advancing the dredge the proper distance while keeping on course, without overswinging.

If two or more line-of-sight references are being used, and they are aligned when the dredge is on the correct compass heading, the leverman knows his dredge is on the correct course. His responsibility is to advance the dredge the proper amount, swing to both sides of the channel, while cleaning up the bottom to the correct depth, all the while remaining on course.

SWING ANGLE FOR ADVANCE

Assume we have a dredge with a hull length of 70 feet and a distance from the spud to the cutter of 100 feet. Assume further that there is a distance between spuds of 10 feet, a hull width of 30 feet, and a channel width requirement of 220 feet. The material being dug requires an advance of 5 feet for each set. See Fig. 21-1. With the walking spud arrangement shown, we must calculate the advance swing angle at which to start the walking action of the dredge. The forward movement of the walking or starboard spud, when swinging on the port spud, equals the sine of the swing angle times the distance between spuds. One half of the desired 5-foot advance is achieved by the walking spud; therefore,

$$\text{Sine } \theta = \frac{2.5}{10} = .25$$

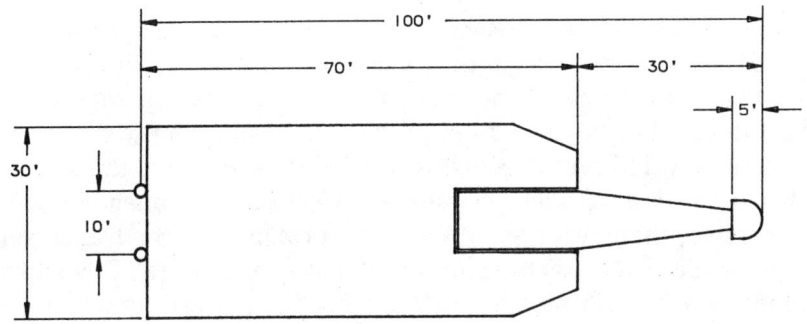

Fig. 21-1. Dredge plan view for swing analysis.

Advance Angle θ equals 14.5° as obtained from our hand-held calculator. The starboard spud is dropped when the dredge has swung to the port 14.5°. Then the port spud is raised, and the dredge is swung 14.5° past the course centerline to the starboard. Here the port spud is lowered and the starboard spud is raised, and the dredge has advanced 5 feet on course.

SWING WIDTH

With the maximum swing angle allowable for the dredge of 45° (in practice it is less), the maximum swing in one direction would be:

Sine 45° × 100 feet = 70.7 feet

Swinging the same distance in the other direction gives a maximum channel width dug in one pass of 141.4 feet. Since the required channel is 220 feet, obviously two passes must be made. By planning a 10-foot overlap in the middle of the passes, the dredge should be programmed for 115 feet each pass, or 57.5 feet each side of the dredge centerline. To calculate the operating swing angle:

$$\text{Sine } \theta = \frac{57.5}{100} = .575$$

Swing Angle θ = 35°

ANCHOR LOCATION

Adhering to the 45°-angle limitation of swing wire to dredge centerline (to avoid spud overload), we lay out the 35°-swing angles calculated above, and then the maximum 45° angle of swing wire. See Fig. 21-2. The swing wire projections outline the limits of the acceptable locations of the swing anchors. By scaling, we see that with the anchor 300 feet from the channel centerline, we have approximately 115 feet acceptable advance distance for the anchor. This represents 23 advance sets of 5 feet each. If anchor booms were used, their reach would allow approximately 6 advance sets to be made before the anchor would have to relocate. This shows almost a 4 to 1 advantage for the anchor barge over booms, which partially explains many operators' preference for the barge.

If instead of 35°-swing angles, 40° were used, the dotted lines on Fig. 21-2 would apply. This shows that at 300 feet from channel centerline, only 9 advances would be available before the anchors would have to be relocated and about 2 if anchor booms were used. While viable, this is not as convenient and efficient as the 35°-swing angle. On the other hand, the channel width with the 40° swing is increased by approximately 14 feet over the 35° swing, and if the width were needed, could quite possibly justify the more frequent anchor movement. For most dredges 35° is a good, functional swing angle (70° inclusive). Seldom should an angle of 40° be exceeded on a 45°-design-basis dredge. The reason for this becomes obvious when using 45°-swing angles (also shown in Fig.

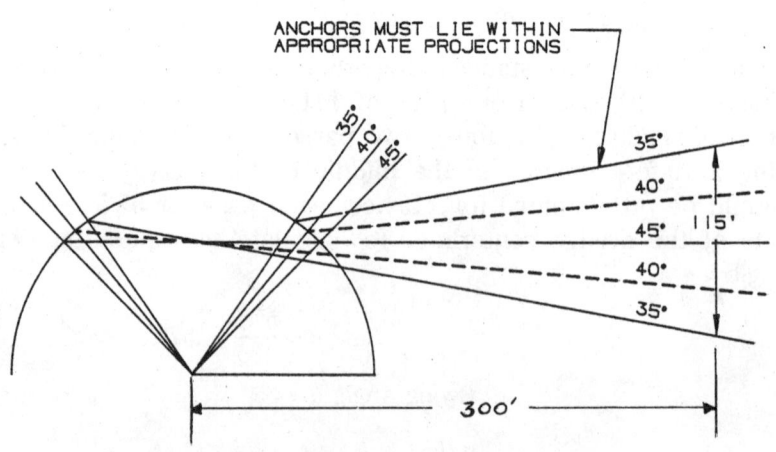

Fig. 21-2. Swing angles vs. anchor positions.

OPERATION AND TROUBLESHOOTING

21-2), we find that the anchor line projections superimpose one on the other leaving no room for error in anchor location or advancement.

Spuds can, of course, be designed to a more rigorous standard, e.g., for 50° or 60° anchor wire angles to the hull. This is expensive, and a dubious expenditure, since extending the swing angle past 45° has the serious shortcoming of reducing the material available to the cutter as the swing angle increases. As Fig. 21-3 shows, for every foot of cutter advance on the channel centerline, there is only 81.9 percent of that foot at a swing angle of 35°, 70.7 percent at 45°, and zero at 90°.

CHANNEL WIDTH LIMITATIONS

Channel width capability increases with swing angle, but obviously a point of diminishing returns is reached long before zero material is available at 90°. Generally a 30°- to 40°-swing angle (60° to 80°

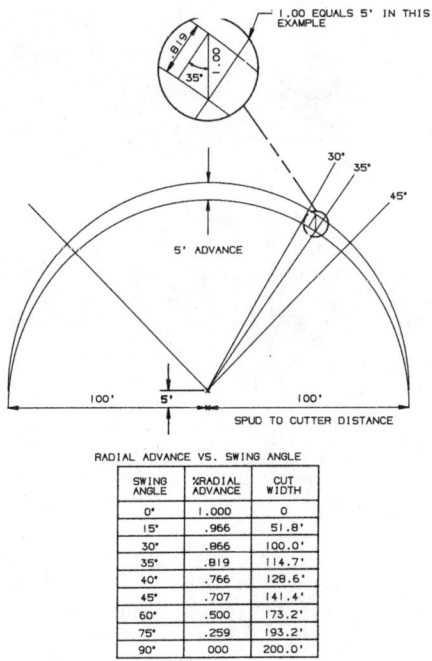

Fig. 21-3. Radial advance vs. swing angle.

inclusive) is satisfactory. This leads to a convenient rule of thumb which states that the channel width achievable by a dredge is 100 to 130 percent of its length from spud to cutter, when the ladder is in digging position. Fig. 21-3.

There is also a minimum swing angle when the dredge is excavating a body of water with an initial depth less than the draft of the dredge. Fig. 21-4 shows that for the exemplary dredge, the swing angle must be a minimum of 24° (48° inclusive) in order for the cutter to create adequate width for the hull. To avoid this problem with conventional dredges, canal dredges have been developed with swinging ladders mounted on trunnions at the bow of a nonswinging hull.

The cropping of the corners of the bow of the dredge is to alleviate the problem of minimum channel width, as well as to avoid contact between the swing wire and the hull. Wires have reputedly worn holes in the hull in this fashion.

The 24° minimum and 40° practical maximum swing angles point out a design limitation which should be considered on all cutterhead dredges. This dredge has a minimum swing width of 90

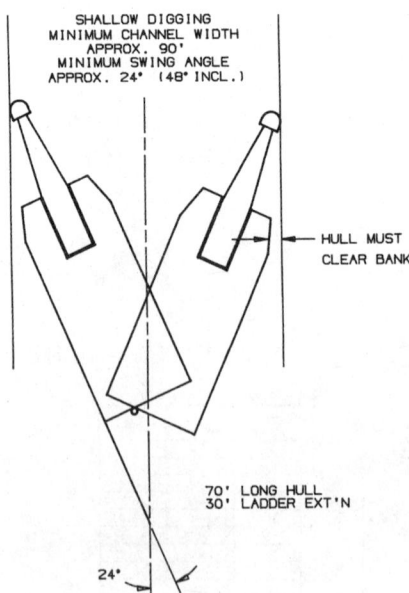

Fig. 21-4. Minimum channel width.

feet (in shallow digging only) and a practical maximum of 130 feet. This relatively narrow range can be improved by lengthening the ladder with respect to the hull. A practical ratio of hull length to ladder extension beyond the hull is the 7 to 3 ratio of our example, but a longer ladder (lower ratio) gives greater flexibility.

TROUBLESHOOTING

A dredge owner-operator complained of the problem of negligible production after two days of operation. He described the conditions as a high vacuum, and a low discharge pressure. Since the material being dug was not difficult, he was confident the cutter was not the problem.

When told that the symptoms were classic for an obstruction in the suction line, his reply was the equivalent of "you know that, and I know that, but this dredge doesn't seem to know that. Please come take a look at it."

Inspection of the dredge confirmed that his vacuum was a steady 24 inches, and his pressure only a fraction of what it should be. Upon shutdown, there was no obstruction in the pump cleanout, and a rod detected no problem when thrust down the suction line. The rod was thrust up the suction inlet through the cutter, still detecting nothing. Since high vacuum existed indicating high losses in the suction, and low discharge pressure indicating low losses in the discharge line, there was a disparity in apparent flows between the suction and discharge. While these symptoms could have occurred with high flow in the suction line with accompanying high flow in a shortened (broken) discharge line, inspection disclosed no break. Therefore, the problem was necessarily an obstruction in the suction line which caused low flow in both suction and discharge. Since there was no foreign object in the suction line, we examined the flexible suction hose at the trunnions, and discovered the hose had collapsed to a small, flat opening. Upon removing the hose and correcting its inside diameter, we were able to get the dredge to work normally.

Logic and understanding the pumping system should allow the dredge operator to diagnose most problems in minutes. Among the most difficult to diagnose are problems with the suction hose, e.g., a flap torn loose from the inner lining of the hose under high vac-

Fig. 21-5. 10-inch lightweight, portable dredge. Courtesy: Ellicott Machine Corp.

uum conditions, but which swing back into place under low vacuum. The problem cost one operator much more than minutes of his time. Logic eventually prevailed and he isolated the hose as the problem, and its removal for inspection disclosed the cause.

ABNORMAL GAUGE READINGS AND THEIR MEANING

Following are some troubleshooting symptoms and their probable causes. It should be kept in mind that vacuum indicates losses in the suction line, while pressure indicates losses in the discharge line.

High Vacuum—Low Pressure
This condition, described above, suggests an obstruction in the suction line or a broken discharge line. There are other instruments

which can supplement the information provided by the vacuum and pressure gauges. The horsepower indicator on the pump drive would rise significantly if the discharge line was broken, since horsepower rises with flow rate. Further, since the engine would probably lug down with the high load, the RPM indicator would suggest high flow. If a production meter existed, low flow and low production would be an immediate indication of suction line obstruction.

High Vacuum—High Pressure
This condition is not necessarily an abnormality. If high production is shown on the production meter, or reported at the disposal area, the operator should continue. If production is low, it could indicate an obstruction in both suction and discharge lines. This is relatively rare, but can occur in trashy conditions.

Low Vacuum—High Pressure
Obviously this condition indicates low flow through the suction, so that the high pressure is indicative of an obstruction in the discharge line. On very long lines, this may be normal, since the operator will have determined that to raise the vacuum by picking up more solids, would overload his pumping system. A production meter would be very helpful here to help control specific gravity.

Low Vacuum—Low Pressure
This condition is normally associated with inadequate pick-up of solids. The cause could be inadequate advance, improper positioning of the cutter, or material too hard or too trashy for the cutter. The condition would also occur if the pump were running too slowly.

Noise Change
Sound or noise is a surprisingly good operational indicator to the experienced leverman, and it is not limited to the dredge pump or other major equipment. The writer observed a leverman who detected by listening that his service water pump had shut down. It is true that he had a service water pressure gauge on his control panel, but this gauge merely confirmed what his auditory senses had already told him.

Excessive Swing Force
Unless accompanied by high cutter force, excessive swing force generally means the ladder is dragging. The cause could be poor positioning of the cutter or a bank cave-in.

Excessive Cutting Force
This condition can be caused by excessive swing speed, excessive advance or lowering, or difficult material. In hard material it is sometimes advisable to raise the cutter on the return swing to avoid having the cutter "run away." The return swing under these conditions should be accomplished at maximum swing speed to conserve time.

SUMMARY

This chapter outlines some serious but common hazards of the dredge. Operational procedures are explained to avoid these problems. Also, troubleshooting techniques for the slurry pumping system are covered.

Chapter 22

THE ENVIRONMENT AND THE DREDGE

The dredging industry has been in the middle of the great national debate which has raged for years in regard to the environment. This debate has been clouded by the emotions of two groups of well-intentioned people.

ENVIRONMENTALISTS VS. DEVELOPERS

The first group includes those who have observed the process of deterioration of our environment by the excesses of industrial and commercial development and have dedicated themselves to halting the process. They sometimes appear to operate on the premise that everything natural is good, and everything man-made is bad.

The second group includes those who see the needs of the world's burgeoning population and dedicate themselves to supplying the homes, the automobiles, the minerals, etc., required by mankind. They sometimes give the impression that any disruption to nature is justifiable to supply man's needs.

To dispel the cloud of emotion engendered by this debate, it is necessary to establish the essential facts and definitions which will allow both groups to achieve a common perspective, and thus reach agreement on appropriate policies, objectives, and procedures. This is not easy.

WATER POLLUTION DEFINED

The following definition is suggested: *Water pollution is any change which affects the practical condition of the water adversely for any purpose.* While this definition may seem logical, inclusive, and rigorous enough to please all, probably neither side would endorse it enthusiastically (which perhaps makes it about right). The developers may well say that progress requires some sacrifice in environmental standards. On the other hand the environmentalists may say that they must also guard against *suspected* adverse affects which may not show up for years. The author, who considers himself both an environmentalist and developer, recognizes the considerable merit in both views.

Any body of water has a degree of assimilative capacity of pollutants, just as the atmosphere can assimilate CO_2 expelled from the lungs of man. Therefore, addition of pollutants which can be assimilated in a body of water without the practical condition of the water being affected adversely would not constitute pollution by the above definition.

POLITICS AND PUBLIC OPINION

There is no doubt that public opinion and the legislative pendulum have swung toward the environmentalist's view in the last two decades. It is unfortunate that laws have been passed without adequate knowledge, justification, or proper evaluation of their effects. Not only has money been wasted, but some expenditures have been counterproductive to their intent. Because of the extreme definition of wetlands, worthy projects have been denied, and owners have been deprived of their property rights. These problems are being recognized in today's debates, and the legislative pendulum shows signs it may be starting to reverse. Hopefully, this time, we can stop the pendulum near the midpoint, where the interests of mankind and the environment are properly served. Both are worthy causes.

TURBIDITY

Most dredge people consider that among the most unreasonable regulations are those that refer to turbidity. They recognize that at its worst, the dredge is like a grain of sand on the beach when compared to the turbidity effect of a spring rain, e.g., on the Mississippi, which conveys two million tons per day of silt into the Gulf of Mexico. A moderate wind over a shallow bay or lake waters can stir up turbidity that the dredge could never equal. Millions upon millions of dollars have been spent on dredged material deposition areas to provide retention and settling time to reduce the turbidity of the returning water to, say, 50 Jackson Turbidity Units above background. This test of light transmission through an effluent sample may be justified in some cases where the receiving waters are extraordinarily sensitive, but in most waters which have proven to have amazing recuperative powers, the test is probably useful largely in exercising the authority of the governing body. Dredge people must learn to live with this, however, and Waterways Experiment Station Technical Report DS-78-10 *Guidelines For Designing, Operating and Managing Dredged Material Containment Areas*[13] is highly recommended.

DREDGE AS CLEANUP TOOL

The dredgeman and the environmentalist are not natural adversaries, although some members of both groups seem to operate on the premise that they are. Actually, to clean up a polluted body of water, the dredge is a very effective tool. There are numerous examples of dredge work in: river cleanup; beach nourishment; reservoir and lake restoration; creation of bird sanctuaries; flood control; creation of man-serving projects (e.g., Panama Canal, New Jersey Turnpike, Washington National Airport, etc.); and many others. Although the dredge is a disinterested, objective tool, its use on unwise projects has led to antipathy toward the entire dredging industry.

EFFICIENCY VS. ENVIRONMENTAL DISRUPTION

Perhaps the most important contribution the dredge operator can make is to operate his equipment in the most efficient manner. Instead of operating at a solids content of 10 percent, he should operate at least at 20 percent. Note that the water per cubic yard of material transported is reduced by more than 55 percent when the solids content is doubled. A strong case can be made that *environmental disruption is inversely proportional to the efficiency of the dredging operation*. This puts a burden on the operator that he should gladly bear, since it contributes directly to more favorable economic results.

Fig. 22-1. Horizontal auger dredge for light-duty pond cleaning. Courtesy: Ellicott Machine Corp.

RECOMMENDATIONS

The justification of any project is achieved after the facts have been gathered and judgment rendered as to whether or not the advantages of the project outweigh the disadvantages. But unfortunately, the criteria by which the project is judged are widely divergent depending upon the group doing the judging. The Audubon Society, the Izaak Walton League, or the Sierra Club have their environmental views, which emphasize the value of pristine, natural sites, untouched by pollution; the industrialist and the developer have their views which are more closely allied with the economy, human needs, and convenience. In the final analysis, these criteria must merge since no thinking industrialist or developer wants to eliminate worthwhile flora or fauna, any more than the environmentalist chooses to live in a cave rather than cut the trees or mine the copper required for a decent residence.

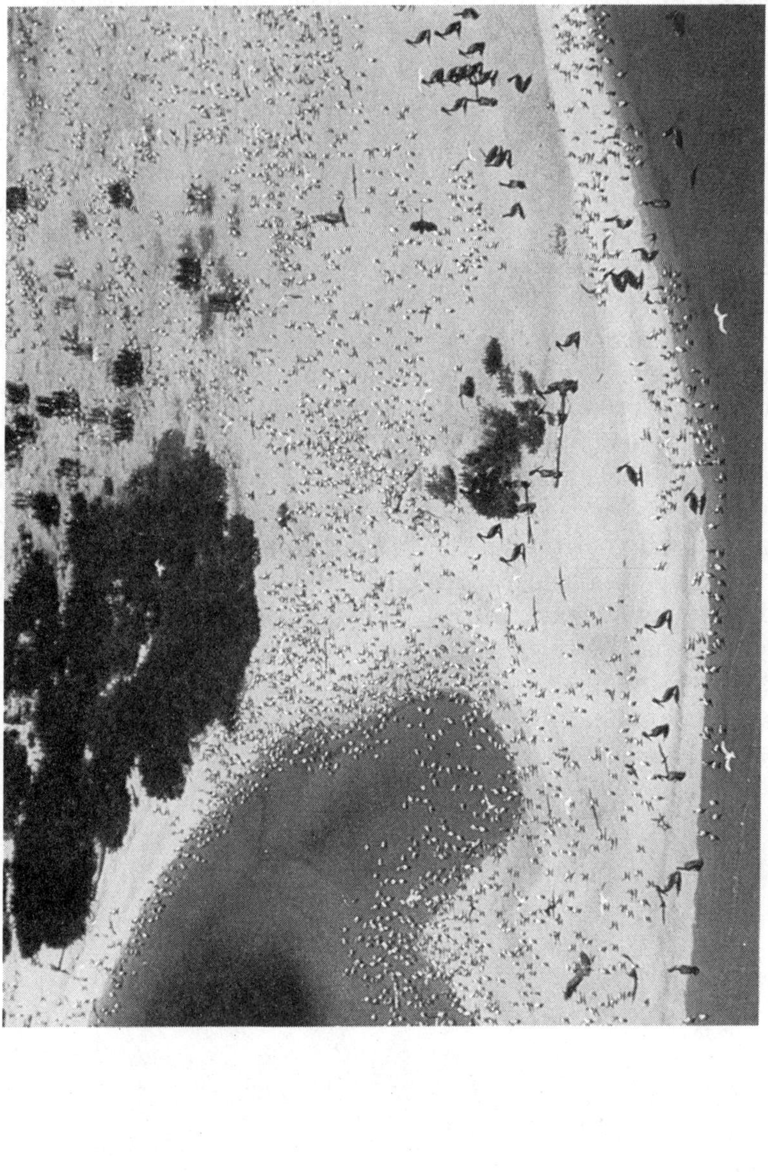

THE ENVIRONMENT AND THE DREDGE 243

Fig. 22-2. Bird sanctuary on dredge disposal area. From U.S. Army Corps of Engineers publication, *Dredging is for the Birds*.

It is essential that all developers recognize the intrinsic value of nature's attributes and attempt to disrupt them to the minimal extent compatible with man's well-being. On the other hand, the environmentalist must recognize that man has a right to exist and inevitably affects the natural scheme of things. By walking through the woods, man damages plant life and exhales carbon dioxide into the air. He destroys ecological microcosms as he clears a forest to plant corn or wheat. Environmental disruption in the final analysis is a function of population. A case can be made that the environmentalist's efforts would be more productive if directed toward the world's population explosion, rather than at the minor details of a dredging project.

The answer does not lie in the cessation or deferral of development projects which afford mankind the products and conveniences he demands; rather, we must engage in the necessary national dialogue that brings together the conflicting viewpoints, requiring the use of the most efficient technology and allowing us to arrive at the correct political judgment after reflecting upon the economic, sociological, and philosophical aspects. This is not an easy process. We cannot begin too early to sweep away the emotionalism and get on with the rational judgments that are so essential to the well-being of mankind and his society. Each segment of the community has a responsibility to assist in this process, and the dredgeman has a great stake in its success. He must recognize that the environmental disruption of a hydraulic dredge has an inverse relationship to its operational efficiency; therefore, it is incumbent upon him to operate in the most efficient manner. The author sincerely hopes this book can contribute to that end.

ABBREVIATIONS

A	Area
As	Area of suction pipe
A_D	Area of discharge pipe
C	Friction coefficient for Hazen & Williams equation
cu ft or ft^3	Cubic feet
m^3	Cubic meters
cu yd or y^3	Cubic yards
cu yd/hr	Cubic yards per hour
C_{HP}	Horsepower coefficient
C_v	Solids fraction by true volume
Che	Head-efficiency coefficient
D or d	Diameter in inches
D_d	Inside diameter of discharge pipe
DD	Digging depth
DE	Dredge efficiency
d_{50}	Median grain size of soil
D L	Dredge law
D_s	Inside diameter of suction pipe
f100	Friction loss per 100 feet of pipe in feet of water
F	Coefficient of friction for Darcy-Weisbach equation
ft/sec	Feet per second
g	Acceleration of gravity (32.2 ft/sec^2)
G or g	Grams
gal	Gallon
g/cc	Grams per cubic centimeter
g/l	Grams per liter
GPM or q	Gallons per minute
H or h	Head in feet of liquid
H_b	Head, barometric

H_e	Head, entrance loss
H_f	Head, friction loss
HG	Mercury
HP	Horsepower
hr	Hour
HSG	Head, specific gravity (to lift solids)
H_v	Head, velocity
ID	Inside diameter
L_s	Suction lift (water surface to pump centerline, + or −)
LL	Line length, discharge
l/s	Liters per second
m or M	Meters
m/min	Meters per minute
m/s	Meters per second
N or RPM	Revolutions per minute
OD	Outside diameter
lbs	Pounds
lbs/cu ft	Pounds per cubic foot
psi	Pounds per square inch
psig	Pounds per square inch gauge
q or GPM	Gallons per minute
RPM or N	Revolutions per minute
S or s	Seconds
SG	Specific gravity
SG_s	Specific gravity of slurry
SG_w	Specific gravity of water
V	Velocity in feet per second
V_d	Velocity in discharge pipe
V_s	Velocity in suction pipe
W	Weight per cubic foot

USEFUL FORMULAS AND CONVERSION FACTORS

Area of sphere	$= 12.566 \times R^2$
cu yd/hr	$= D^2 \times \text{Vel} \times (\text{Avg SG-1}) \times .661$
F	$= 1.09 \times (100/C)^{1.85} \times V^{1.85} / D^{1.1655}$
GPM in pipe	$= 2.448 \times D^2 \times \text{Vel}$
Head (by water pump)	$= C_h \times (D \times N/1{,}840)^2$ (D is impeller diameter)
Head (by water pump)	$= C_h \times V^2 / 2g$ (V is impeller tip speed)
HP (pump)	$= \text{GPM} \times H_s / (3{,}960 \times E_s)$
HP (pump)	$= \text{GPM} \times H_w \times \text{SG}/(3{,}960 \times E_w)$
HP (pump)	$= V \times D^2 \times H_s/(1{,}616 \times E_s)$
HP (pump)	$= V \times D^2 \times H_w \times \text{SG}/(1{,}616 \times E_w)$
HP	$= T \times N/5{,}252$
SG	$= 1.1 \times \text{Volume\%} + 1$
Torque	$= 5{,}252 \times \text{HP}/N$
Vel (peripheral)	$= D \times N/229.2$
Vel (in pipe)	$= \text{GPM}/(2.448 \times D^2)$
Volume% (slurry)	$= (\text{SGs-1})/1.1$
Volume of sphere	$= 4.189 \times R^3$
1 cubic meter per hour	$= 4.4$ GPM
1 liter per second	$= 3.6$ m^3/hr $= 15.85$ GPM
1 meter	$= 39.37$ inches $= 3.2808$ feet

LEGEND

C	= friction factor, Hazen-Williams;
C_h	= coefficient of head for pump (value greater than one);

D	= diameter in in.;
E_s	= pump efficiency on slurry;
E_w	= pump efficiency on water;
F	= friction loss in feet of water per 100 ft of pipe;
g	= acceleration due to gravity;
HP	= horsepower;
H_s	= slurry head (in feet of water);
H_w	= water head (in feet of water);
N	= RPM;
R	= radius;
SGs	= specific gravity of slurry;
T	= torque in lb ft;
V or Vel	= velocity in ft/ sec;
Volume %	= percentage of in situ sand in slurry.

REFERENCES

1. Turner, Thomas M. (1970). "The basic dredge laws." *Proceedings, World Dredging Conference*, 411–424.
2. Huston, John (1970). *Hydraulic dredging*. Cornell Maritime Press, Cambridge, Md.
3. Herbich, John B. (1975). *Coastal and deep ocean dredging*. Gulf Publishing Co., Houston, Tex.
4. *Cameron Hydraulic Data*. (1984). Westaway and Loomis, eds., Ingersoll Rand.
5. Basco, David (1975). "Pump design affects performance." *World Dredging and Marine Construction*, (Jan.), 10–12.
6. Cornet, R. (1975). "Wear in dredgers." *The Dock and Harbor Authority*, LVI(655).
7. Pokrovskaya, V. N. "Means of increasing the effectiveness of hydrotransport." (Notes from technical paper in author's file.)
8. Taylor, E. W. (1948). "Micro hardness testing of metals." *Institute of Metals*, Vol. 74.
9. Wellinger, K., and Vetz, H. (1955). "Sliding, scouring and blasting wear under influence of granular solids." VDI Forschungsheft, 21 B(1955), 449.
10. Van Den Haak, V. (1972), "Anchors." *Holland Shipbuilding*, (Oct.), 14 A.
11. U.S. Army Corps of Engineers. (1978). "Prediction and control of dredges material dispersion around dredging and open-water pipeline disposal operations." *Waterways Experiment Station Technical Report DS-78-13*, Washington, D.C.
12. Terry, Leland E. (1967). "It's what's up front that counts." *Proceedings, World Dredging Conference*, 91–113.
13. U.S. Army Corps of Engineers. (1978). "Guidelines for designing, operating, and managing dredged material containment areas." *Waterways Experiment Station Technical Report DS-78-10*, Washington, D.C.
14. Cave, I. (1976). "Effects of suspended solids on the performance of centrifugal pumps." *Hydrotransport*, 4(May), 18–21.
15. U.S. Army Corps of Engineers. (1978). "Confined disposal of dredged material." *Engineering Manual 1110-2-5027*, Washington D.C.

PUBLIC LAW 95-269—APR. 26, 1978

92 STAT. 218

PUBLIC LAW 95-269
95TH CONGRESS

An Act

Apr. 26, 1978
(H.R. 7744)

To amend the acts of August 11, 1888, and March 2, 1919, pertaining to carrying out projects for improvements of rivers and harbors only contract or otherwise, and for other purposes.

Be it enacted by the Senate and House of Representatives of the United States of America in Congress assembled.

Rivers and harbors, improvements.

That section 3 of the Act of August 11, 1888 (25 Stat. 423; 33 U.S.C. 622), is amended to read as follows:

"Sec. 3. (a) The Secretary of the Army, acting through the Chief of Engineers (hereinafter referred to as the "Secretary"), in carrying out projects for improvement of rivers and harbors (other than surveys, estimates, and gagings) shall, by contract or otherwise, carry out such work in the manner most economical and advantageous to the United States. The Secretary shall have dredging and related work done by contract if he determines private industry has the capability to do such work and it can be done at reasonable prices and in a timely manner. During the four-year period which begins on the date of enactment of this subsection, the Secretary may limit the application of the second sentence of this subsection for work for which the federally owned fleet is available to achieve an orderly transition to full implementation of this subsection.

"(b) As private industry reasonably demonstrates its capability under subsection (a) to perform the work done by the federally owned fleet, at reasonable prices and in a timely manner, the federally owned fleet shall be reduced in an orderly manner, as determined by the Secretary, by retirement of plant. To carry out emergency and national defense work the Secretary shall retain only the minimum federally owned fleet capable of performing such work and he may exempt from the provisions of this section such amount of work as he determines to be reasonably necessary to keep such fleet fully operational, as determined by the Secretary, after the minimum fleet requirements have been determined. Notwithstanding the preceding sentence, in carrying out the reduction of the federally owned fleet, the Secretary may retain so much of the federally owned fleet as he determines necessary, for so long as he determines necessary, to insure the capability of the Federal Government and private industry together to carry out projects for improvements of rivers and harbors. For the purpose of making the determination required by the preceding sentence the Secretary shall not exempt any work from the requirements of this section. The minimum federally owned fleet shall be maintained to technologically modern and efficient standards including replacement as necessary. The Secretary is authorized and directed to

Study undertake a study to determine the minimum federally owned fleet required to perform emergency and national defense work.

Submittal to Congress The study, which shall be submitted to Congress within two years after enactment of this subsection, shall also include preservation of employee rights of persons

PUBLIC LAW 95-269—APR. 26, 1978

92 STAT. 219

presently employed on the existing federally owned fleet. "Sec. 2. Section 8 of the Act of March 2, 1919 (40 Stat. 1290; 33 U.S.C. 624), is amended to read as follows:
"Sec. 8. (a) No works of river and harbor improvement shall be done by private contract—

"(1) if the Secretary of the Army, acting through the Chief of Engineers, determines that Government plant is reasonably available to perform the subject work and the contract price for doing the work is more than 25 per centum in excess of the estimated comparable cost of doing the work by Government plant; or

"(2) in any other circumstance where the Secretary of the Army, acting through the Chief of Engineers, determines that the contract price is more than 25 per centum in excess of what he determines to be a fair and reasonable estimated cost of a well-equipped contractor doing the work.

"(b) In estimating the comparable cost of doing the work under sub-section (a) (1) by Government plant the Secretary of the Army, acting through the Chief of Engineers shall, in addition to the cost of labor and materials, take into account proper charges for depreciation of plant, all supervising and overhead expenses, interest on the capital invested in the Government plant (but the rate of interest shall not exceed the maximum prevailing rate being paid by the United States on current issues of bonds or other evidences of indebtedness) and such other Government expenses and charges as the Chief of Engineers determines to be appropriate.

"(c) In determining a fair and reasonable estimated cost of doing work by privated contract under subsection (a) (2), the Secretary of the Army acting through the Chief of Engineers, shall, in addition to the cost of labor and materials, take into account proper charges for depreciation of plant, all expenses for supervision, overhead, work-men's compensation, general liability insurance, taxes (state and local), interest on capital invested in plant, and such other expenses and charges the Secretary of the Army, acting through the Chief of Engineers, determines to be appropriate".

Approved April 26, 1978

LEGISLATIVE HISTORY:

HOUSE REPORT No. 95-605 (Comm. on Public Works and Transportation)
SENATE REPORT No. 95-722 (Comm. on Environmental and Public Works)
CONGRESSIONAL RECORD:
    Vol. 123 (1977): Sept. 27, considered and passed House.
    Vol. 124 (1978): Apr. 5, considered and passed Senate, amended.
                 Apr. 13, House agreed to Senate amendments.

INDEX

Abbreviations: 245
Acceleration of gravity: 5
Advance, dredge: description, 26; diagram, 27; mechanisms, 213; swing angle, 230
Affinity laws, pump: 9
Altitude, effect on: HP, 45; suction velocity, 44
Anchors: booms for, 189; calculating weight, 189; positioning, 231; types, 187, 188
Atmospheric pressure: see Barometric head
Automation: dredge operation, 205; velocity control, 203
Bank height: 29, 213
Barometric head effect on: flow, 5; production, 43
Basket cutter: calculations, 124; capacity chart, 128; configuration, 117, 118, 119, 120; cutting force, 122; description, 113; drives, 122; HP, 123; materials of construction, 127; particle passage, 128; speed, 123; torque, 121
Bidding: chapter 19, 208
Booster pump: see Pump, booster

Bucket wheel: advantages, 130; description, 128; disadvantages, 131
Calculating the project: chapter 19; use of computer, 220
Cavitation: chapter 11; chart, 95; definition, 94; eye speed effect, 96; pump speed effect, 96
Channel width limitations: 233
Charts: cavitation limits, 95; coefficient head efficiency, 12; cutter capacity, 128 ; dredge cycle, 83; dredge efficiency, 29; Hazen-Williams friction, 91; production, chapter 8; velocity limitation, 35
Coefficients: head-efficiency chart, 12; HP, 62, 63, 64
Computer: chapter 20, 220; advantages, 227; need, 220; software, 221, training, 226; database 221; production charts, 223, 224, 225
Costs, estimating: chapter 19, 208
Cutters: capacity chart, 128; endless chain, 132; high speed disc, 133; see also Basket cutter and Bucket wheel

Diesel engine: dredge pump compatibility, 150; torque curve, 150
Digging depth, effect on: production, 43; suction velocity, 53; bidding, 211
Discharge line size, effect on: friction, 58; line length, 60; production, 61, 71, 76; HP, 59
Dredge configurations: D dredge, 68, 70; with ladder pump, 72, 78; with booster pump, 71, 77; with ladder and booster pumps, 72, 79; L dredge, 73, 79
Dredge cycle: chart, 83; description, 82, 84
Dredge efficiency: chart, 29; definition, 26; effect on the environment, 241
Dredge Laws: I, 15; II, 25; III, 32; IV. 41; V, 49; VI, 56; VII, 69
Dredge types: compensated cutterhead, 108; cutterhead, 107; plain suction, 12; trailing suction (hopper), 102
Drives: cutter, 122; dredge pump, 151; ladder pump, 165; winch, 186
Efficiency: dredge, 26; dredge chart, 29; drives, 149; pump, 12
Engine, diesel: dredge pump compatibility, 150; torque curve, 150
Environment vs. the dredge: 239
Entrance loss: definition, 51; equation, 51
Flow regimes: 87

Formulas and conversion factors: 247
Friction loss: C coefficient, 90; C chart, 91; calculation, 89; Fanning equation, 59; Hazen-Williams equation, 91; particle size, 93; pipe size, 58; prediction of losses, 92
Gas vs. the dredge pump: 165
Grain size: median, 36; classification systems, 36
Gravitational force: 4
Head losses: entrance, 51; friction, 52, 90; lift, 52; specific gravity, 52; terminal elevation, 89, 211; velocity, 50
Hopper dredge: 102
Horsepower, pump: coefficient, 62, 63, 64; equation, 57; effect on GPM, SG, head, 57; line length, 56; recommended, 65; vs. gpm and pipe diameter, 59
Hydraulic dredge: definition, 3
Impeller: 141; entrance, 141; eye, 147; geometry, 98, 148; mounting, 142; shrouds, 141; tip speed, 148; wiper vanes, 143
In situ volume: 16, 19
Instruments and controls: chapter 18, 199; dredge automation, 205; production meter, 203; velocity control, 203; velocity gauge, 205
Ladder pump: see Pump, ladder
Life, wear: see chapter 16, 173; equation, 174; hardness, 175; particle size, 178; velocity, 177; wear zones, 183

INDEX 257

Lift head: 52
Limiting velocity chart: 35
Line length: HP relationship, 56
Line size: effect on HP, 62
Newtonian fluid: 7
Operational errors: 229
Particle clearance, pump: 138
Particle size vs. C factor: 93
Percent solids: maximum, 25
Pollution defined: 239
Porosity of soils: percent voids, 17; weight, effect on, 17
Production: charts, 68, 70, 71, 72, 73, 80; equations, 19; meter, 203; percent solids effect, 11; suction size effect, 60
Project calculation: chapter 18, 208; advancing mechanism, 213; bid price, 218; calculation method, 209; calendar time, 216; contract documents, 209; costs, 217; cutter, 212; digging depth, 211; dredge efficiency, 214; line length, 212; production rate, 214; production time, 216; soil type, 210; swing width, 213; suction size, 214; terminal elevation, 211; total yardage, 215; trash, 217; work face height, 213
Public Law 95-269: 250
Pump, booster: coordination with dredge pump, 167; effect on line length, 167; location, 169; effect on production, 78; production charts, 72, 73, 80

Pump, centrifugal: performance curve, 9, 10; principle, 8
Pump, dredge: adjustable mounting, 143; casing, 146; drives, 149; Che chart, 12; elevation effect, 53; eye speed, 147; eye to diameter ratio, 148; head-efficiency coefficient, 12, 149; HP coefficients, 62, 63, 64; impeller, 141; particle clearance, 138; shaft and bearings, 142; slurry effect, 11; speed vs. HP, 9; SG effect, 11; stuffing box, 142; tip speed, 148; torsional vibration, 152; thrust, 153; wear lining, 139; wiper vanes, 143
Pump efficiency: Che chart, 12; effect of slurry, 11
Pump, ladder: design requirements, 161; drives, 165; production effect, 78; design, 78, 79
Sand-water mixture characteristics: table, 18; chart, slurry, 20
Slurry, effect on hydraulics: 3, 11
Slurry, volume to weight conversion: 21; grams per liter conversion, 22
Soil: expansion factor, 16, 21; classification chart, 37; types, 90
Solids percent: average, 15, 26; chart, 20; factors affecting, 26; in situ volume, 16; maximum percent volume, 25; ta-

ble sand-water mixture, 18; true volume, 19
Solids percent: table of sand-water characteristics, 18
Specific gravity: effect on pump, 11
Specific gravity head: definition, 52; equation, 52
Spuds: carriage for, 194; design, 193; purpose, 193; type lift, 193, 194
Spreadsheet examples: 223, 224
Suction lift: 52
Suction line: inlet, 115; jet booster, 162; optimum size, 61; size effect on production, 61; table of values, 61
Suction line losses: velocity head, 49; entrance, 51; friction, 52; SG head, 52; lift, 52
Suction line size, effect on: bidding, 214; discharge line wear, 76; production, 61; pumping distance, 60
Suction velocity: optimum, 53; table for 24-inch dredge, 53; computer calculation, 55
Swing width: 29, 213, 231
Table: sand-water mixture characteristics, 18
Temperature effect: on diesel HP, 45; air density, 46
Terminal elevation, 211
Test, standard penetration: 125, 128
Thrust, pump: 153
Torsional vibration: 152

Training: by computer, 226
Trash: 217
Troubleshooting: 235; abnormal gauge readings, 236; excessive swing force, 238; excessive cutting force, 238; noise, 237
Turbidity: 240
Turbulence, slurry :hydraulic transport effect, 32; requirements for different materials, 33
Velocity head: definition, 3; equation, 4
Velocity, limiting: chart, 35; pipe size effect, 34
Velocity, suction: altitude effect, 44; optimum vs. digging depth, 50, 53; pipe size effect, 34; optimum vs. solids percent, 53
Volume, soils: in situ, 16, 19; true, 19; chart, 20; table, 18
Water hammer: 167
Water pollution defined: 239
Wear: see Life, wear
Weight, sand and mixtures: table, 18; slurry chart, 20
Winches: anchor booms, 189; brakes, 187; fleet angles, 187; swing and ladder, 185; line pull, 186; spud hoists, 192; swing speed, 186
Wiper vanes: 143, 145
Wire rope: design limitations, 198; type, 194
Work face height: 29, 213